The Physics
of
Mr. Tweed

Gary Petersen

Contents

Introduction

"The truth of our faith becomes a matter of ridicule among the infidels if any Catholic, not gifted with the necessary scientific learning, presents as dogma what scientific scrutiny shows to be false."
<u>St. Thomas Aquinas</u>

Mr. Tweed.   Hello. I am Mr. Tweed and I will help you through a short broad survey of physics.

Kay.   Since we are only spending one semester on this subject, I understand what you mean by short. What do you mean by broad?

Mr. Tweed.   Because this may be your only college exposure to science, the scope of the material should be large enough to introduce the major elements of physics in order to provide you with a broad overview that you can integrate into the other aspects of your education. Therefore, this course is designed to include classical mechanics, electricity & magnetism, waves and light, relativity, quantum mechanics, nuclear physics, astronomy, cosmology, and a discussion of unsolved questions in physics today.

Jill.   How are we going to do all that in one semester? My perception of physics is that it is a very difficult subject that must be studied for some time in order gain any significant comprehension.

Mr. Tweed.   I don't think any of you aspires to become a scientist. However, you are all bright and hardworking and capable of appreciating, in physics, some of the most fun weird stuff that you will ever encounter.

In order to produce a continuous understandable development of ideas during the limited time available, every aspect of the subject which is unnecessary to the overall plan will be dropped. This means many practical topics will be missing. You will concentrate on fundamental basics.

Jill. You think we are bright but I am not even sure what physics is about.

Mr. Tweed.   When God made the universe, he did more than just make the heaven, earth, sun, moon, and the stars. He created the rules for how all the matter in this universe interacts. We have a name for these rules; they are called the *laws of physics*. It is the task of the physicist to discover these basic laws and put them into some sort of useful form that can be understood by not only scientists but by the rest of us as well. When I say find these laws, I don't mean dig up some tablets with the laws carved in some ancient language. The scientist must search, examine, hypothesize, and test experimentally, if possible, every weird idea that comes along. He must pore through huge amounts of data and attempt to reduce this information into principles. If he is very gifted and clever, he might synthesize some of these principles into a law.

It seems to me that, if we love God, we should appreciate his creation in more than just an aesthetic way. We should push our intellectual capability as far as we are able to see its beauty and grandeur in every way possible. Having an expanded view of God's creation is awe inspiring, beautiful, and mind blowing.

I know you can do it and have fun at the same time. Fritz will help you.

Kay.   Who is Fritz?

Mr. Tweed.   We have a machine to simulate physical phenomena and Fritz is the assistant inside the machine.

Jill. Will there be a lot of mathematics?

Mr. Tweed.   Physics is married to mathematics and there is no way to really separate them without trivializing the subject.   However, every effort has been made to minimize the mathematical difficulties.

I apologize for leaving out the names of many great physicists.

Prerequisites

The prerequisites are:

1)   Simple algebra including some skill with ratio and proportion
2)   Scientific notation
3)   Right angle triangle trigonometry
4)   Simple geometry
5)   Simple graphs
6)   Use of a scientific calculator

See you in the first class.

Session 1

Forces and what they do

Everybody continues in its state of rest, or of uniform motion in a straight line, unless it is compelled to change that state by forces impressed upon it.
--Isaac Newton

Mr. T.   Let us begin by examining what you know about force. Jill, what does a force on an object do?

J.   It causes that object to move.

Mr. T.   Good! Kate, can you describe the motion?

K.   As I understand it, it is not so much just motion but a change in motion. If the object is initially at rest, the force will make it move in the direction of the force. If it is already moving, the force will cause it to speed up, slow down or change direction. The force causes the motion to change at a certain rate.

Mr. T.   Excellent! Kate, what is the name for this motional rate of change?

K.   I don't remember.

Mr. T.   It is called acceleration and we will get into it in a few minutes. Before we do, however, let me ask the question: There are forces on you right now. Why are you not changing your motion (accelerating)?

J.   I don't understand what forces are acting on me.

Mr. T.   Well, the force of gravity is one, and it is pulling you down. Why are you not accelerating in a downward direction?

J.   The chair keeps me from moving.

Mr. T.   The chair exerts an upward force that exactly counteracts gravity. We can say that the resultant force on you is zero.

J.   What do you mean by resultant force?

Mr. T. If you add all the forces acting on you, the result is zero. The resultant force is the sum of all the forces. If the resultant was non-zero you would be accelerating in the direction of the resultant. This addition process is a little tricky because forces have magnitude and direction. They are vectors and we must use vector rules when adding them.

J.   But the acceleration has to depend on more than the force; a big object will accelerate at a lower rate than a small one for the same force. I can apply enough force to accelerate a baseball but not an automobile.

Mr. T. You are correct, but it is not the size that is important; it is something called the mass. Also, in the case of an automobile, you would have to overcome the force of friction in order that the resultant force would be non-zero.

There is a simple equation from **Isaac Newton** that relates the quantities resultant force ($\vec{F}$), mass (M), and acceleration ($\vec{a}$).

$$\vec{F} = M\vec{a} \qquad (1.1)$$

Why did I put arrows above the F and a?

K.   Because they are vectors and have both magnitude and direction.

J.   Can you explain exactly what you mean by acceleration? I thought it meant to go faster and faster.

Mr. T. It could mean in one case to go faster and faster, it could also mean in another case to go slower and slower. Actually, it is defined as the rate of change of velocity. The following formula relating time (t), velocity ($\vec{v}$), and acceleration ($\vec{a}$) may make it more clear.

$$\vec{a} = \Delta\vec{v} / \Delta t \qquad (1.2)$$

The Δ (delta) symbol refers to change in. Thus $\Delta\vec{v}$ is the change in velocity and Δt is the change in time or time interval involved.

J.   I am not sure what all this means.

Mr. T. It is time for some experiments. We have a new simulator and you are the first to use it. Please take one of the helmets and put it on.

J.   But it will completely cover my head.

Mr. T.   You will be able to hear, see, and feel the simulation once the helmet is on.

J.   Ok, here goes.

Inside the simulator

F.   Welcome

J.   Who was that?

F.   I am Fritz your friendly computer simulator assistant. I am here to help you set up the simulations, answer questions, and make measurements.

K.   Look, Jill, there is a large automobile. What is that for? Fritz, are we going for a ride?

F.   No! Jill mentioned that she could not accelerate a car by pushing on it. In the real world, she could not overcome the force of friction. However, here in the simulator, we can make the friction between the car and the road zero. We can also make the friction between your feet and the road very high so that she or you will not slide and fall on your face when you push. Put on the friction boots and gloves. Without the gloves, your hands would just slide off the car.

Jill, how much force do you think you can exert when you push?

J,   I have no idea. What are the units of force? What is a large force for someone like me?

F,   The unit of force is a Newton. The equation (1.1), $\vec{F}=M\vec{a}$ works when the units of mass, distance, and time are respectively kilograms (Kg), meters (m), and seconds (s). The units of velocity are meters/second or $\frac{m}{s}$ and since $\vec{a}=\Delta\vec{v}/\Delta t$, the units of acceleration are $\frac{\frac{m}{s}}{s}$ or $\frac{m}{s^2}$. Actually, the unit of force could be given as $\frac{Kg\,m}{s^2}$. However, it is easier to say Newtons.

Because your mass is about 60 *kg*, the force of gravity on you (your weight) is approximately 600 *Newtons*. It is possible but difficult for you to exert forces greater than your weight. However, you can easily exert a force equal to a small fraction of your weight.

J.   Fritz, that's all very complicated, but since you think my weight is about 600 *N*, I think I can easily push with a force of 60 *N*. How long should I push?

F.   Why don't you try for one full minute (60 *seconds*)? The simulator will help you push for exactly that amount of time and with a force of 60 *N*.

J.   OK, here goes.

Look the car is actually moving. I quit pushing but it continues to move. How fast is it going?

F.   There is no friction so the speed is now a constant 1.8 *m/s*.

Kate, can you tell me the mass of the car?

K.   Well, the velocity changed from zero to 1.8 *m/s*, so $\Delta\vec{v}=1.8$ *m/s*, and the time interval was 60 *seconds*, so $\Delta t = 60s$. The acceleration can be found from equation (1.2), $(\vec{a}=\Delta\vec{v}/\Delta t)$.

$\vec{a}=\frac{1.8}{60}=3.00\times 10^{-2}\frac{m}{s^2}$.   The mass can be found from equation (1.1) $(\vec{F}=M\vec{a})$. $M=\vec{F}/\vec{a}=\frac{60}{3.00\times10^{-2}}=2.00\times 10^3 Kg$

F,   Good job! Let's try something else and use some of you knowledge about vectors. Let's fix it so that the car is frictionless in the rolling direction but not in the sideways (right angle) direction. The tires will only roll and not slide sideways. Try pushing sideways.

J.   I can't because the car is gone. I started it rolling and it never stopped. We can't even see it now.

F.   This is only a simulation. I will bring it back.

J.   Wow, that's cool, Thanks! Now I will stand next to the door and push on the side.

It doesn't move. It just pushes back.

F. Try pushing at some angle other than 90 *degrees* from the rolling direction.

J. What angle would you like?

F. I am just a simulator; I don't care.

J. I will try 45 *degrees*, the same force (60 *N*), and for the same time (60 *seconds*).

    The car is moving but not as fast as before.

F. Yes, it is moving 1.27 *m/s* in the rolling direction.

J. How do I make sense out of this?

K. We have to draw a vector diagram of your 60 *N* force, its component in the rolling direction, and its sideways component.

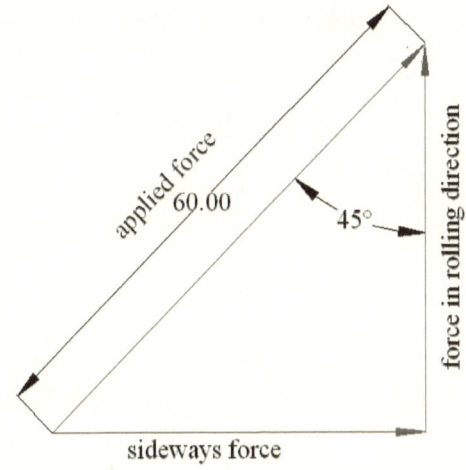

Fig. 1.1

The components of Jill's push on the car

J. I don't understand vectors very well but now that you have drawn the diagram, I think I can solve for the components. We have a right triangle and the hypotenuse is 60 N, so we can use trigonometry to find the sides.

Force in the rolling direction = $60 \times \cos45° = 42.4\ N$

Force in the sideways direction = $60 \times \sin45° = 42.4\ N$

The sideways force will be cancelled by the frictional forces between the car tires and the road. The force component in the rolling direction will accelerate the 2000 *Kg* car. The acceleration should be:

$$\vec{a} = \frac{\overrightarrow{F_{rolling}}}{M} = \frac{42.4}{2 \times 10^3} = 2.12 \times 10^{-2}\frac{m}{s^2}$$

If $\Delta t{=}60s$, the change in velocity can be found from equation (1.2),

$\Delta\vec{v} = \vec{a} \times \Delta t = 1.27 m/s$. This is the same number you gave us.

F.   Are you ready for something just a little more involved?

J&K.   Don't make it too hard!

F.   Suppose you both push at different angles. The car is pointed north so we can use that as a reference for the angles.

K.   I will push a little harder than Jill with a force of 80 *N* and I will direct the force 30 degrees to the west of North.

J.   I will push with 60 *N* directed 60 *degrees* East of North.

K.   The car is moving fairly fast to the North. How fast is it going?

F.   2.98 *m/s*. Can you make sense out of it and calculate its speed?

K.   Because of the sideways friction, we only have to calculate the North components of our forces and add them to find the resultant. Dividing by the mass should give the acceleration, and finally multiplying by the time interval will give the velocity.

$$F_{North} = 80\cos(30°) + 60\cos(60°)$$

$$= 69.3 + 30 = 99.3\ N$$

$$\Delta\vec{v} = \frac{\vec{F}}{M} \times\ \Delta t = \frac{99.3 \times 60}{2 \times 10^3} = 2.98\ m/s$$

Fritz, this is the same number you gave us so I got it right.

F.   We have one more simulation in this session which involves adding all the components of the vectors. I will give you an object which can move in any direction on a frictionless surface. As before your shoes will have friction so that you can push on the object. The simulator will help you to push with the force and direction you designate.

J.   Look! It's a rock that's much bigger than we are.

K.   Let's push with the same forces, directions, and time that we did with the car. This time, we can do the calculations first, determine the resultant force magnitude and direction and velocity, push and see if we get it right. What is the mass of the stone? We will need it.

F.   The mass is 1750 *Kg* (1.75×10^3 *Kg*) or about 30 times your mass.

J.   Kay, please show us how to do it

K.   We have to add our forces. Probably, the easiest way to do this is to draw a vector representing your force and calculate the components in the North and East directions. We will do the same thing with my force and then add the North component of my force to the north component of yours. Because the components are in the same direction, simple addition is all that is needed. The next step is to add the East-West components. The results give the two components of the resultant force.

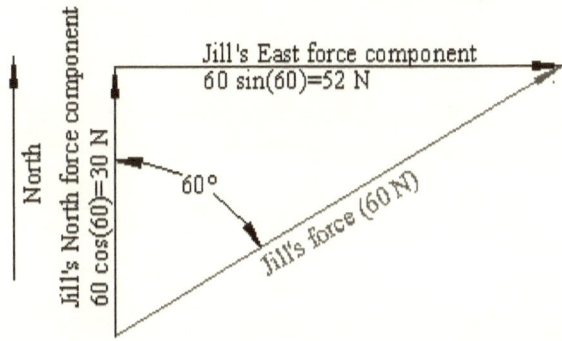

Fig. 1.2
Jill's force on the rock

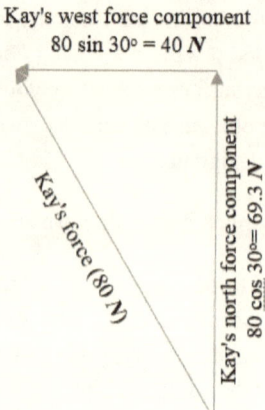

Kay's west force component
80 sin 30° = 40 *N*

Kay's north force component
80 cos 30°= 69.3 *N*

Kay's force (80 *N*)

Fig 1.3
Kays force on the rock

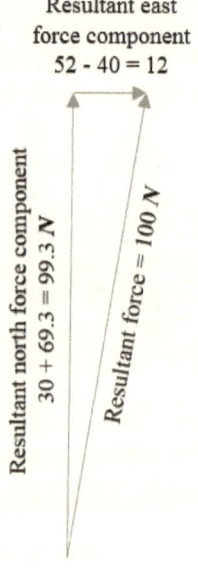

Resultant east
force component
52 - 40 = 12

Resultant north force component
30 + 69.3 = 99.3 *N*

Resultant force = 100 *N*

Fig 1.4
The resultant of both Kay's and Jill's force

$$F_{North} = 69.3 + 30 = 99.3 \ N$$
$$F_{East} = 52 - 40 = 12 \ N$$

The magnitude of the resultant can be found from the Pythagorean Theorem.

$$F = \sqrt{(99.3^2 + 12^2)} = 100 \ N$$

The angle is found using trigonometry:

$$angle = tan^{-1}\left(\frac{12}{99.3}\right) = 6.89 \ degrees \ \text{East of North}$$

The acceleration is:

$$\vec{a} = \frac{\vec{F}}{M} = \frac{100}{1.75 \times 10^3} = 5.71 \times 10^{-2} \frac{m}{s^2}$$

The velocity:

$$\Delta\vec{v} = \vec{a} \times \Delta t = 5.71 \times 10^{-2} \times 60 = 3.43 \frac{m}{s}$$

K&J.   Let's push and see if we got it right.

The direction seems ok and it is moving fairly fast. Fritz, did we get it right?

F.   You did! Congratulations and this ends this simulation. Take off your helmets.

Back in the classroom

Mr. T.   Did you enjoy yourselves?

J.   We did! However, I am not yet very comfortable with vectors.

Mr. T.   We will use them again and again, and you will have a chance to get better. In fact, we have one more topic for this session and it also involves vectors. We need to understand the concept of momentum.

J.   I understand the word and its general use. Does it have a different meaning in physics?

Mr. T.   Not much! Something with lots of momentum will be difficult to stop. Clearly, anything with high velocity or great mass will be hard to stop. The physics definition of momentum is simple. The momentum ($\vec{p}$) is equal to the mass (M) times the velocity ($\vec{v}$).

$$\vec{p} = M\vec{v} \tag{1.3}$$

K.   That seems simple enough. What can we use it for?

Mr. T.   The law of conservation of momentum is one of the most useful concepts in physics. It states that the momentum of a system (the vector sum of the momentums of all its parts) can never change unless the system is subjected to an outside force.

K.   I don't understand exactly what that means. Could you please explain it?

Mr. T.   I think it's time for another simulation. You will examine a simple system that contains only two objects. There will be a collision between the objects and the situation will be made even simpler because the collision will take place all along a straight line, no angles to worry about. Maybe Fritz can get the concept across. Please put your helmets on.

Inside the simulator

F.   Welcome back! The objects you see today will feel no gravity or friction with their surroundings; they will, however, be subject to friction with each other. We need a sign convention for this experiment. Things to the right will be positive and to the left will be negative.

K.   What is that big light colored blob hanging in space?

F.   If you look to the left you will see a smaller 1 Kg dark blob moving with a velocity of 2.4 m/s toward the larger 2 Kg stationary light colored blob. Watch and see what happens when there is a collision.

J.   The dark blob was absorbed and the two of them are moving together to the right with a slower velocity. How fast are they moving?

F.   0.8 m/s. The situation is shown in figure 5

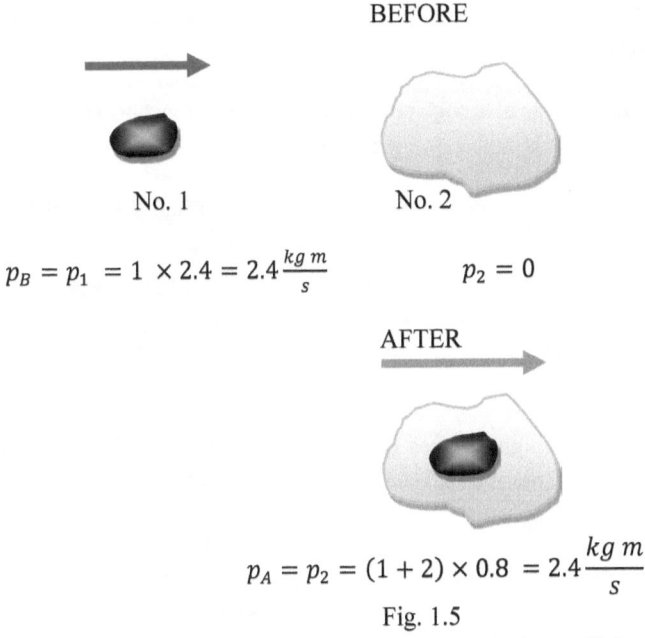

BEFORE

No. 1   No. 2

$$p_B = p_1 = 1 \times 2.4 = 2.4 \frac{kg\ m}{s} \qquad\qquad p_2 = 0$$

AFTER

$$p_A = p_2 = (1 + 2) \times 0.8 = 2.4 \frac{kg\ m}{s}$$

Fig. 1.5

Blob momentum before and after collision

Notice that I have used subscripts on the quantities to keep track of what and when. p_1 stands for the momentum of the dark blob (No. 1) before collision.

J.   I get it. Then, p_2 stands for the initial momentum of the light colored blob (No. 2) and it is zero because it is not moving. p_A must stand for the total momentum after the collision. p_B would stand for the total momentum before and is equal to p_A because the dark blob was the only one moving.

F.   Good, now what conclusions can you draw from your results?

K.   The final momentum was the same as the initial momentum. The total momentum did not change. Momentum was lost by the dark blob and gained by the light colored one.

F.   Excellent! Your first conclusion comes about because Newton tells us that for any action there is an equal and opposite reaction. The force on the dark blob by the light colored one is equal and opposite to the force on the light colored blob by the dark one and the two forces both act for the same amount of time. You can show that $\vec{f} \Delta t = M \Delta \vec{v} = \Delta \vec{p}$. $\vec{f} \Delta t$ is called the impulse and it tells us what the change in momentum is. The dark blob receives an impulse to the left (-) and the light colored one to the right (+).

In this situation momentum was conserved (did not change) and there were no outside forces.

J. I understand, but that was a strange sort of collision; we don't normally see blobs colliding or for that matter doing anything.

F. That was an example of a completely inelastic collision and the simulator made it very simple. That's the value of a simulation; we can do very simple ideal things. We have to learn about kinetic energy in order to define exactly what an elastic collision is but for now we can simulate one. We will keep the mass of each object the same as before but instead of blobs we will use hard balls.

J. I see the stationary 2 *Kg* light colored ball and here comes the 1 Kg dark ball at 2.4 *m/s*. The dark ball has the same velocity as the dark blob in the previous example.

Bang! They hit but now the dark ball rebounded to the left and only the light colored ball is moving to the right. What are their velocities now?

F. The dark ball has a velocity of -0.8 *m/s* and the light colored velocity is 1.6 *m/s*.

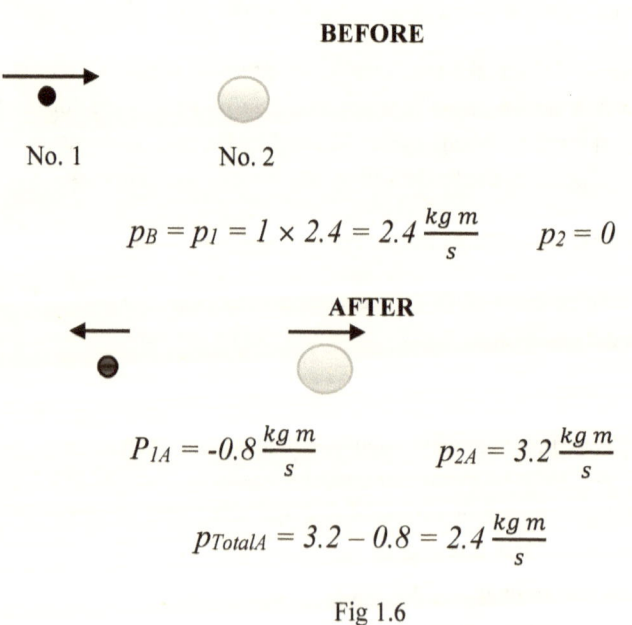

BEFORE

No. 1   No. 2

$$p_B = p_1 = 1 \times 2.4 = 2.4 \frac{kg\ m}{s} \qquad p_2 = 0$$

AFTER

$$P_{1A} = -0.8 \frac{kg\ m}{s} \qquad p_{2A} = 3.2 \frac{kg\ m}{s}$$

$$p_{TotalA} = 3.2 - 0.8 = 2.4 \frac{kg\ m}{s}$$

Fig 1.6
Before and after ball collision

F.   Jill, can you analyze these results?

J.   I will try.

1.   The change in the momentum of the dark (No. 1) ball (-3.2) was exactly opposite the change in momentum of the light (No. 2) colored ball (+3.2).
2.   The final momentum (2.4) was the same as the initial momentum (2.4). The total momentum did not change.
3.   The elastic collision resulted in greater speeds for both balls than in the blob collision.

F.   Good job! You have shown that momentum is conserved in this case. We have not investigated cases where both balls are moving or when things happen in 2 or 3 dimensions; we will return to these concepts later. This ends the simulation and this lesson. Take off your helmets.

Session 2

Surface Gravity, Energy, and Gas

Gases are distinguished from other forms of matter, not only by their power of indefinite expansion so as to fill any vessel, however large, and by the great effect heat has in dilating them, but by the uniformity and simplicity of the laws which regulate these changes. --James Clerk Maxwell

Mr. T.   We begin this lesson by examining some aspects of gravity on the surface of our planet, trying to understand the concepts of kinetic and potential energy, and finally applying almost everything we have learned so far to the kinetic theory of gas.

J.   Are we going to do all that today?

Mr. T.   I hope so. Galileo showed that all objects fall at the surface of Earth with the same acceleration. From what we learned last time, this means that the force of gravity must depend directly on the mass of the object. $F = gM$, where g refers to the acceleration due to gravity and has a value of -9.8 $\frac{m}{s^2}$. The negative sign indicates that the acceleration is down. Notice that I have decided and not to put the arrows above the F and g. They are still vectors and must be treated as such.

This is very interesting because it shows that the mass has two functions; it shows how much inertia an object has and also what the gravitational force will be. Let's go to the simulator and try some things.

In the simulator

F.   I see that you have returned; welcome back!

K.   I see a weight marked 20 *kg,* a large one meter high table, and a second table next to it that is two meters high. What are we supposed to do?

F.   Jill, lift the weight onto the lower table.

J. OK

It's heavy; that's a lot of work.

F. Exactly! Now, Kay, stand on the first table and lift the weight to the second.

K. I hope I don't fall.

F. Don't worry; this is a simulator. Did you do more or less work than Jill?

K. I started higher, but otherwise I lifted it the same distance. I think I did the same amount of work as she did.

F. We can draw some conclusions from this. Twice the amount of work was required to raise the weight 2 meters as 1 meter. The amount of work also depends on the force you exert (force of gravity = -Mg). We can express this mathematically:

$$Work = U_g = Wt. \times height = -Mgh \qquad (2.1)$$

Where g is the acceleration of gravity (-9.8 m/s^2) and h is the height (displacement) the object was raised. The sign convention may get a little confusing at this point. The force of gravity is down (-) ; the acceleration of gravity is also down, but the force you exerted on the weight was up and the displacement was up (+). Note that if you had picked the weight up only a millimeter and moved it across the room and sat it down, you would have done no work at all because the force must be in the same direction as the displacement. They are both vectors.

J. I understand the equation, but what is the U_g term used for?

F. The symbol U is used for potential energy. In this case all the work you did went into gravitational potential energy.

K. Are you saying that work and energy are the same thing?

F. Work refers to the transfer of energy from one form to another. It has the same units as energy which are:

$$\frac{kgm^2}{s^2} \ or \ Newton \ meters \ (Nm) \ or \ Joules \ (J)$$

J.   You said work was the transfer of energy from one form to another. I transferred energy to the weight. Where did this energy come from?

F.   It was stored as chemical potential energy in your body. The chemical energy was released causing your muscles to contract as you lifted the weight. The net result was that you changed chemical potential energy into gravitational potential energy.

Let's try something different. Would you like to play catch? I have here a baseball which has a mass of 0.145 *kg* (145 *g*). How fast do you think you can throw it? A professional baseball pitcher can throw about 40 *m/s* (90 *miles/hour*).

J.   I think I can throw it about ¼ as fast as the pitcher or about 10 *m/s*.

F.   I know this is hard but I want you to throw it straight up with that velocity, but before you do I want Kay to tell me about how much time it will take to reach the top of its flight.

K.   I think that is easy. The velocity at the top is zero, so the change in velocity is -10 *m/s* . The acceleration is - 9.8 *m/s²* which means that the change in velocity (Δv) in each second would be -9.8 *m/s,* or about -10 *m/s*. Therefore it should take about one second to reach the top.

F.   Jill, try it.

J.   It took about 2 seconds to come back to my hand, one second up and one second down. Should the time to come down be the same as the up time?

F.   Yes, the situation coming down is just reversed from going up; the initial velocity is zero and the final velocity will be -10 *m/s* which gives a change of -10 *m/s*. This is the same change as before. Kay, can you figure out how high (*h*) the ball goes?

K.   This is a little bit difficult because the ball is changing speed. The velocity starts as 10 *m/s* and ends at zero, but because the change is uniform with respect to time, I think we can use the average velocity which is $\frac{10+0}{2} = 5$ *m/s*. The correct time to reach the top can be found from *g* $=\Delta v/\Delta t$. $\Delta t = -10/-9.8$ = 1.02 *s*. The height is, therefore, (average velocity × Δt) or 5 × 1.02 = 5.1m *or* 16.7 *ft*.

F. Correct! Suppose the baseball pitcher threw the ball straight up with four times the velocity would it go four times as high?

J. First of all, I don't think the pitcher could throw that hard straight up but if he did I think it would go 4 times as high.

F. Jill, do the calculations and see if you are right.

J. OK! $\Delta t = -40/-9.8 = 4.08\ s$, which is 4 times as long. However the average velocity is $\frac{40+0}{2} = 20\ m/s$ which is also 4 times as big. Therefore, the distance will be 4 times 4 or 16 times as high ($4.08s \times 20m/s = 81.6\ m$). That is really high! I guess I was wrong.

F. You were wrong but you figured it out. I am proud of you. Kay, I want you to stand on top of the two meter table. You should be able to drop the ball from a distance of 3 meters above the floor.

K. It fell, hit the floor, bounced up about one meter, fell again, and bounced up a few centimeters. It took less than a second to hit the floor.

F. I am going to ask for something a little more difficult now. I want you to figure out the mathematical expression for h in terms of its velocity when it first hit the floor.

J. I think we can do it but we are probably never going to calculate the distance a ball falls from its final velocity. Why do we want to do this?

F. Mathematical expressions are designed to give us insights into the physical world. It is said that "a picture is worth a thousand words". A similar statement could be made about a mathematical equation. After you find the expression, with only a little more work you will see the insight it can provide.

J. OK. Here goes. If the ball starts with zero velocity and ends up with v_f, the average velocity will be $v_f/2$. The time required to reach v_f should be: $\Delta t = v_f/g$. Therefore,

$$h = average\ velocity \times \Delta t \quad \text{or}$$

$$h = \frac{v_f^2}{2g} \qquad\qquad (2.2)$$

F. I like it. If we multiply both sides of this equation by *Mg* and use the expression for Ug in equation (2.1), we get

$$U_g = Mgh = M\frac{v_f^2}{2} \qquad (2.3)$$

This is an important relation because it shows us that when potential energy is converted to motion, the energy of the motion (kinetic energy) is equal to one half the mass times the velocity squared. In general the kinetic energy of an object is

$$KE = M v^2 / 2 \qquad (2.4)$$

Thus when we raised the ball (mass = .145 *kg*) 3 meters, we gave it some gravitational potential energy which we can calculate as $(0.145 \times 9.8 \times 3)$ *or* 4.26 *J* . When the ball fell its potential energy was converted to kinetic energy.

In this simulation the ball bounced several times but each time it did not go as high as the last. This means the collision was not completely elastic. Some of the energy seemed to be lost each time it bounced. If the ball had been completely inelastic, like the light colored and red blobs, it would have hit the floor and stopped.

K. Where did all the energy go when we dropped the original ball and it quit bouncing?

F. The energy went into heat! The energy of motion of the whole ball was converted to random motion of its molecules.

K. How can we tell if that's true?

F. You could measure the temperature of the ball before and after it was bounced. You would need a very accurate thermometer, but it could be done.

Let's make the ball completely elastic (better than a super ball). Kay, drop it again.

K. Wow! It fell and bounced all the way back to my hand and now it just keeps repeating the process. Will it ever stop?

F. Only when the simulation is terminated. Let's examine the situation:

1. The ball starts with gravitational potential energy but zero kinetic energy ($v=0$).
2. As the potential energy is lost the ball speeds up and the kinetic energy is increased.
3. When the ball hits the floor, the direction of the velocity ($-7.67 m/s$) reverses but the kinetic energy remains unchanged. The square of a positive or negative number is positive.
4. As the ball rises it loses kinetic energy which is converted to potential energy.
5. When it reaches the top, the whole thing starts over.

K. We are finished with this simulation but in the next one you will encounter some perfectly elastic collisions. See you next time.

Back in the classroom

Mr. T. Hi! I have something more challenging for you now. We will use almost everything you have learned to this point. It is called the kinetic theory of gases. We will try to relate the kinetic energy of gas molecules to the pressure of a gas. The pressure (P) is the force per area on a surface in the gas. Our atmospheric pressure at the earth's surface (1 atmosphere) is about 15 *psi (pounds per square inch)* or 1×10^5 N/m^2. As you go higher above the surface this pressure decreases. At an altitude of 20 miles (32 *km*) it is nearly zero. The theory we develop will be very simple but it works pretty well until the pressure or temperature get too high. Most of the work will be in the simulator.

You will be transported to a cubic room having a distance of only 1×10^{-8} m on each side. It will, however, appear to be 10 m on a side because you will be very small. The actual size will be one billion (10^9) times smaller than what you see. The simulator will also change the speed of your clock. At this very small size everything will happen so fast that you could not see anything. Your clock will be speeded up by a factor of 100 billion (10^{11}). This means that when you see a time interval of one second, only 10^{-11} seconds have actually elapsed. The side walls of your chamber will also be very heavy and hard so that all collisions with these walls will be elastic.

Don't worry, anything in this simulation will go right through you; you won't get hit. Go for it and have fun!

In the simulator

F. Are you back already? I think your teacher wants me to do all the teaching. What do you think of your room this time?

J. I don't see anything happening. Wait, I see one ball about 14 cm in diameter slowly bouncing straight up and down between the floor and ceiling.

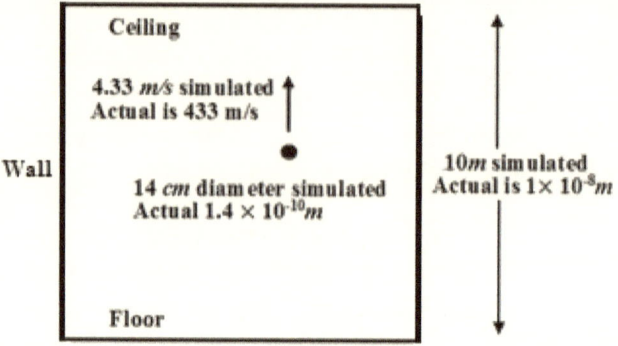

Fig. 2.1
Atom bouncing between floor and ceiling

F. That is a single argon atom moving at about 433 *m/s* which is the average speed it moves at when the temperature is 300°K.

J. What is 300°K?

F. That is normal room temperature on the absolute temperature scale (Kelvin). The degree size is the same as the Celsius scale but zero is the coldest possible temperature. 0°K is the same as -273 °C.

K. As I look at the ball, I am beginning to get the idea. Every time it strikes a wall it changes momentum gives the wall a little impulse. If we average over time all these impulses, we can find an average force.

$$\overline{F} = \frac{Impulse}{time\ between\ impuses}$$

Let's find the time between ceiling strikes.

J. I measure 4.62 *seconds*, but I don't see the point. Gas has molecules moving randomly in all directions striking all the walls. What are we doing looking at one molecule moving up and down and never striking the side walls?

F. Our simulation is very ideal. The motion of the molecule is not random and will never hit the side walls because it hits a perfectly flat ceiling or floor surface at 90° and has no way to lose or gain kinetic energy. I am going to give you a flat reflector panel. Hold it so that the argon molecule hits it and is deflected in such a way that it will strike all the surfaces of the room.

J. OK! Wait a minute; yes it seems to be hitting all the walls, the floor, and the ceiling. It looks random. Is it random?

F. It is not random. It has an x component of velocity (v_x) which can have a positive or negative value but its magnitude must always remain the same. The same is true for the y and z components. However, we can figure out the average forces on the room surfaces and calculate the pressure from this single molecule.

K. Let's find the time between wall strikes.

J. It hits some walls more often than others. I will measure the time between hits on the floor. Kay, measure the time between hits on the ceiling.

K. I get 8 seconds.

J. I got exactly the same and that makes sense because the speed in the up-down (z) direction can't change. I will measure the East wall and you measure the West one.

K. I measured 8.7 seconds. What did you get?

J. I measured the same. What about the North and South walls.

K. The result is 8.06 seconds and I will bet that you measured the same.

The atom has a v_z component that determines the time between hits for the ceiling and the floor. We have to remember that our clocks have been speeded up so the time between ceiling hits is 8×10^{-11} seconds. The distance traveled between hits is twice the size of the cube or 2×10^{-8} m (2b). We must use twice the cube size because the atom must hit the ceiling, travel to the floor, and return to the ceiling for the next hit. Therefore the speed is:

$$v_z = \frac{2 \times 10^{-8}}{8 \times 10^{-11}} \frac{m}{s} = 250 \ m/s.$$

J. I calculated v_x (North-South)= 248 *m/s* and v_y (East-West)= 230 *m/s*

K. We have the velocity components. We can now calculate the magnitude using the Pythagorean Theorem.

$$v^2 = v_x^2 + v_y^2 + v_z^2 \qquad (2.5)$$

J. I did it and the magnitude of the velocity is 433 *m/s*. That's the same number the assistant gave us. So, what do we do now?

K. We have to find the change in momentum when the atom strikes one of the walls. We have found the velocity components but we don't know the mass. The assistant will tell us what the mass is.

F. The molecular weight of argon is 39.9 *amu* and each atomic mass unit (*amu*) is 1.66×10^{-27} *kg*. Therefore, the atomic mass is $39.9 \times 1.66 \times 10^{-27}$ or 6.62 $\times$ 10^{-26} *kg*.

K. We will do the ceiling first. The change in the z component of momentum is twice the momentum the atom has when it hits because it bounces back with the same speed.

$\Delta p = 2mv_z = 2 \times 6.62 \times 10^{-26} \times 250$ or

$$\Delta p = 3.31 \times 10^{-23} \frac{kgm}{s}$$

The average force will be the impulse divided by the time between impulses:

$$\bar{F} = 3.31 \times 10^{-23}/8 \times 10^{-11} = 4.14 \times 10^{-13} N$$

and the average pressure on the ceiling will be the average force divided by the area:

$$\bar{P}_c = \frac{\bar{F}}{A} = \frac{\bar{F}}{b^2} = \frac{4.14 \times 10^{-13}}{1 \times 10^{-16}} \quad or$$

$$\bar{P}_c = 4.14 \times 10^3 \ N/m^2$$

J. The pressure on the floor will be the same as the ceiling. I calculated the East-West pressure as 3.81×10^3 *N/m²*, and the North-South as 4.11×10^3 *N/m²*. When we have only one atom, it is difficult to imagine that the pressure on all the walls would be the same.

F. Work out the formula for the pressure.

K.   I will work it out for the ceiling.

$\Delta t = 2b / v_z$, $\bar{F} = \Delta p / \Delta t$, $\Delta p = 2mv_z$, and $\overline{P_{ceiling}} = \dfrac{\bar{F}}{b^2}$. Therefore,

$$\bar{P}_c = \frac{\bar{F}}{b^2} = \frac{2mv_z}{(\frac{2b}{v_z})b^2} = \frac{mv_z^2}{b^3} = \frac{mv_z^2}{V} \tag{2.6}$$

where V is the volume of the cube.

F.   What do you think will happen if we increase the number of molecules (N) in the cube?

K.   The pressure will be greater and depend directly on the number of atoms (n) and it should be the same on all the walls. Let's try it.

F.   There are now 24 atoms in the room.

J.   Wow, they are all over the place but moving at the same slow place. They are pretty far apart and don't seem to be hitting each other. But look! There is a really slow one but it got hit by a fast one; now both are moving pretty fast. We don't see too many collisions but our clock has been speeded up by a huge factor; I imagine there would be a lot of hits in real time.

I agree with Kate, there appear to be no more hits on one side of the cube than any other. Do we have random motion now?

F.   The fact that we have collisions between molecules will make it random.

K.   This means that $\overline{v_x^2} = \overline{v_y^2} = \overline{v_z^2}$. The bar above the terms indicates an average value. This means, using equation (2.5), that $\overline{v^2} = 3\,\overline{v_z^2}$ or $\overline{v_z^2} = \overline{v^2} / 3$. If we put this into equation (2.6), we ge

$$PV - \frac{m\overline{v^2}}{3} \tag{2.7}$$

I remember that in the first part of this lesson we learned that the $KE = mv^2/2$. If we put this into the equation we get: $PV = \frac{2KE}{3}$. Because the number of hits against the wall (impulses) should follow directly the number of atoms (N) in our sample volume, increasing N should increase the pressure, and the new pressure volume relationship will be:

$$PV = \frac{2N\overline{KE}}{3}$$

(2.8)

F.   Good job! The thing that makes this result so important is that a famous physicist by the name of **Boltzmann** came up with a similar equation:

$$PV = Nk_BT$$

(2.9)

where T is the absolute temperature in degrees Kelvin.

k_B is Boltzmann's constant and is equal to 1.38×10^{-23} J / °K

K.   Let's use this formula to find the pressure:

$$P = \frac{Nk_BT}{V} = \frac{Nk_BT}{b^3} = \frac{24 \times 1.38 \times 10^{-23} \times 300}{(1 \times 10^{-8})^3} = 9.94 \times 10^4 \ N/m^2$$

This is only a tiny bit smaller than the value you gave us (1×10^5 N/m^2) for the atmospheric pressure at the surface of the earth.

F.   Now you understand why I chose those values for the simulator.

K.   Now we can relate the average kinetic energy of the atoms to the temperature.

$$\frac{2\overline{KE}}{3} = k_B T$$

(2.10)

This gives a whole new idea about what absolute temperature is. It is proportional to the kinetic energy of the atoms or molecules at least in a gas. When the kinetic energy is zero (all the motion has stopped), the temperature is zero °K.

F.   Unfortunately, at very low temperatures, the situation becomes more complicated but we won't worry about that now.

J.   What about solids and liquids? We know a little about gas now.

F.   Solids and liquids are more complicated but we can look at the situation a little bit in the simulator. Instead of having massive hard walls to our cube suppose we make it out of copper atoms. Let's look at it.

K.   The walls have an arrangement of about 40 atoms on a side or 1600 atoms on each face. Each atom looks to be about 25 cm in diameter. Unfortunately, they all look very blurry. They seem to be vibrating very fast with an amplitude of about a centimeter.

J.   We have speeded up our clock by factor of 10^{11} and they still look blurry; they must be vibrating with a very high frequency. Look! A slow Argon atom hit the surface and bounced away with much greater velocity than it hit.

F.   The gas atoms are in thermal equilibrium with the copper molecules in the wall of our cube. That means; if they have a lot of kinetic energy and hit the wall, they will transfer some of this energy to the copper atoms. If the Argon atoms have low energy, they will gain energy when they come in contact with the wall and the copper atoms will lose energy.

So far, you have learned about three kinds of energy: Kinetic energy, Potential energy, and the energy of atoms and molecules vibrating. The last type of energy that involves the vibration of atoms and molecules is a form of heat. Before we are finished, we will see a number of situations involving moving atoms, molecules, nuclei, and particles.

This lesson and simulation has come to an end. See you next time.

K.   Good! I am tired of calculating. Good bye!

Session 3

Universal Gravitation and Angular Momentum

A new scientific truth does not triumph by convincing its opponents and making them see the light, but rather because its opponents eventually die, and a new generation grows up that is familiar with it.--Max Planck

Mr. T.   Last time we learned a little about gravity at the earth's surface. Today we want to expand our knowledge of gravity to the solar system, galaxy, and beyond. The first requirement for gaining this understanding of gravity is a measure of knowledge of our solar system. The first person known to have put forth the idea that the earth revolves around the sun and not the reverse was the Greek Aristarchus of Samos (c. 270 BC). His model explained why the planets reversed their direction of motion with respect to the stars as the year proceeded and was simple compared with other models. However the controversy between world views continued even in recent times until 1822 when Pope Pius VII approved a decree allowing the printing of heliocentric (sun centered) books in Rome. The history of the development of this world view is fascinating and worth studying. Unfortunately, we have a short time and must just concentrate on making sense out of the phenomenon (Gravity).

Most of the concepts that were terribly difficult for the thinkers of past ages have been completely internalized by us. We all know that the sun is the center of our solar system. The name "*solar system*" automatically implies that. We also know that the gravitational force from the Sun causes the planets and comets to travel in orbits around the Sun.

What we need to do is to figure out mathematically how this gravitational force changes as we go farther and farther from the Sun. Does the evidence show that this is the same kind of force that holds us on the surface of the Earth? Is this the same force that gives rise to the rings of Saturn? Is it the same force that holds our moon or the moons of Jupiter in their orbits?

All the orbits of planets and moons are actually ellipses and math for elliptical paths is out of our reach. We will, therefore, limit ourselves to orbits like that of Earth that are nearly circular.

The first thing on our agenda is to understand simple circular motion. In order to do this we need to review vector addition. Consider the two vectors shown below. How do we add them?

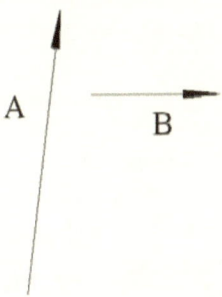

Fig. 3.1
Vectors to be added

K. We move B so that its tail is located at the head of A and the sum is the vector drawn from the tail of A to the head of B.

Mr. T. Correct! Try and see if you can do it.

K. I will move the vector B up to the head of A and draw the resultant.

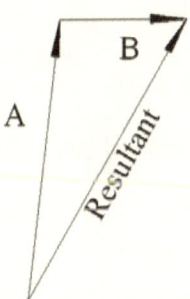

Fig 3.2

Vector addition

Mr. T.   Good! We can read this as $\vec{A} + \vec{B} = \overrightarrow{Resultant}$.

And now, Fritz awaits you. It's time for your helmets.

J.   Are we going to have as many calculations as last time?

Mr. T.   There will be some but not as many as last time.

J.   Good! I am ready. Let's go!

In the simulator

J.   Where am I? It's black in here. Am I alone?

F.   You are alone and on a frictionless surface; try not to fall down. There is a small light on a rope in front of you which is getting closer. When you are close reach down and grab the rope and connect it to your belt. Kate and I can hear you but we can't see you.

J.   I grabbed the rope and have connected myself; I don't want to get lost. The rope is pulling on me but not really hard.

It's been a few minutes now; I must be moving fast but I can't tell because it is dark and frictionless.

F.   Reach into your pocket. You will find a small object similar to a cell phone. Take it out and see if you can turn it on. Don't drop it!

J.   I got it! It has a small screen with compass marks (N, E, S, and W) and an arrow pointing west. Next to the arrow is says 2.7 *m/s*.

K.   It looks like a velocity vector and it must be a GPS device. I wonder if the arrow (vector) gets longer as you go faster.

F.   It is a vector and it will get longer if you go faster.

J.   Let me see, it should be longer. The rope has been pulling for some time now.

It reads the same number (2.7 *m/s*) but it is now pointing South-East. The rope is changing my direction but not my speed. Whoever is pulling the rope must be pretty smart to adjust the pull and direction so that my direction changes and not my speed.

F. Actually, no one is holding the rope. Its end is fixed and the direction the rope makes is at right angles to your motion. You are simply moving in a circle and the rope is fixed at the center of the circle. Because you know your speed, you can determine the radius of the circle by measuring your period (*T*).

J. How can I do that? I don't have a stop-watch and even if I did, I can't see very well with this one little light bulb.

F. Pick a direction and yell start, when you come around to that direction again, yell stop. Kay will measure the time (*T*).

J. Ok, in a moment I will be going north.

Start!

Now I am coming up to north again.

Stop!

K. I read 67.5 *seconds*. The speed is the distance divided by the time and the distance around a circle is: $2 \pi R$. So

$$R = \frac{V \times T}{2\pi} = \frac{2.7 \times 67.5}{2\pi} = 29 \ m$$

J. Wow! I am traveling in a really big circle.

F. Now we need to try figuring out your acceleration. You can feel the resultant force on you, so you must be accelerating. What direction is the acceleration?

J. The acceleration must be in the direction of the force and the only force on me is from the rope. The force must be along the rope and toward the center of the circle.

F. Ok, now in order to find the acceleration we need to see how the velocity vector is changing. Let's look again at your GPS device. Try to look at the velocity once and then a very short time later. We can then figure out the change in velocity and divide by the time to get the acceleration.

J.   The velocity vector is moving around and its tip is making a circle. I can look at it when it is north, wait 1 one hundredth of the way around the circle ($\Delta t = T/100$) and look again. You can see what it looks like below.

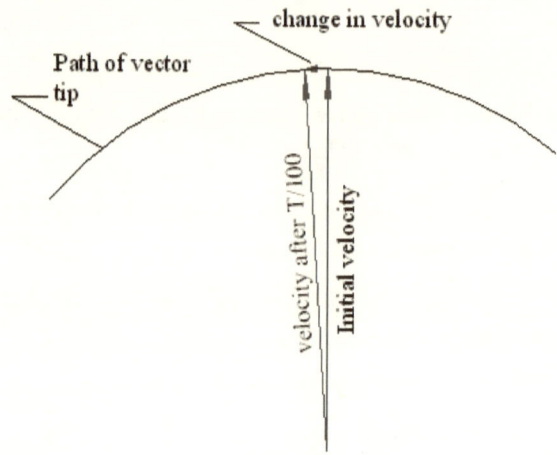

Fig 3-3

Change in Jill's velocity after 1/100 of a period or 0.675 seconds.
Note: The circle shown is not the circular position of Jill, it is a velocity circle.

The change in velocity looks very small and I have no idea how I can measure it.

K.   If we call the circumference of the circle the velocity vector is making C, the change in velocity ($\Delta \vec{v}$) looks to be about $C/100$. Because $\vec{a} = \Delta \vec{v} / \Delta t$, and $\Delta t = T/100$, the acceleration should be simply C / T. The factors of 100 cancel. They could have been 10000, in which case the $\Delta \vec{v}$ would be even closer to the arc of the vector circle.

The circumference of a circle is just 2π times the radius of the vector circle, which in this case is v or 2.7 *m/s*. At this point we have:

$$a = 2\pi v / T \qquad\qquad (3.1)$$

This can be simplified if we recognize that the period T can be expressed in terms of the radius Jill is traveling in and her velocity v. $v = 2\pi R / T$ or $T = 2\pi R / v$. If we substitute these results into equation (3.1), we get two results:

Substituting for v we get:

$$a = \frac{4\pi^2 R}{T^2} \qquad (3.2)$$

and   substituting for T we get:

$$a = \frac{v^2}{R} \qquad (3.3)$$

J.   Why do the a and v symbols not have vector arrows above them? Are they not vectors?

K.   How can you see them, I thought you were in the dark?

J.   I found a button on my little hand held device that lets me see what you are doing. It is amazing what these little devices will do.

K.   OK, I will try to answer your question. The acceleration in this case is always pointing toward the center of the circle in which you are traveling, and the formula only gives its magnitude. The v used here is the velocity component (speed) perpendicular to the rope (radius vector) and for our purposes it is considered to be constant.   There are more complicated possibilities but I don't think we need to consider them now.

F.   That's correct! You have two expressions for the acceleration of an object, in this case Jill, moving with a constant speed in a circle. That was the object of this session, see you next time.

J.   But we didn't find my acceleration. Kay can you calculate it?

K.   I can and I will. Because we know that the radius is 29 m and the speed is 2.7 m/s, we can use equation (3.3).

$$a = \frac{2.7^2}{29} = 0.25 \ m/s^2$$

J.   That's not a lot is it?

K.   Well, the acceleration due to gravity (g) is -9.8 m/s^2, so you are being accelerated only about one fortieth as much as gravity. If gravity were 0.25 m/s^2 and you fell out of a two story window (~10 m), it would take you almost

9 *s* to hit and you would only be moving about 2 *m/s*. This is about the same speed you would reach after jumping off a 25 *cm* stool with our present gravitational acceleration. I think you could handle that.

F.   Alright, you guys, back to class. Goodbye.

In the classroom

Mr. T.   Hi! I hope you have a good understanding of how an object moving with constant speed in a circle can be acceleration. We are now in a good position to understand how the force of gravity changes as we move away from the sun. We can use the planets to examine this situation.

J.   In order to do this we need to know the distance from the Sun to each of the planets. How can we measure these distances?

Mr. T.   It will save us time and effort if use the measurements of Tycho Brahe (1546-1601 AD), and the calculations of Johannes Kepler (1571-1630 AD). Brahe made measurements of the positions of the planets in the sky over a long period of time and Kepler used some very clever geometric and trigonometric techniques to obtain distance data. He was not able to measure the distance from the Sun to the Earth which we call an astronomical unit or *au*. He was, however, able to calculate the distance to the planets in *au* units. Most of the calculations are too involved for us but I can show you how the distance from the Sun to Venus can be done. I will draw a diagram of the Earth, Venus, and the Sun and we will try to figure out the distance from Venus to the Sun.

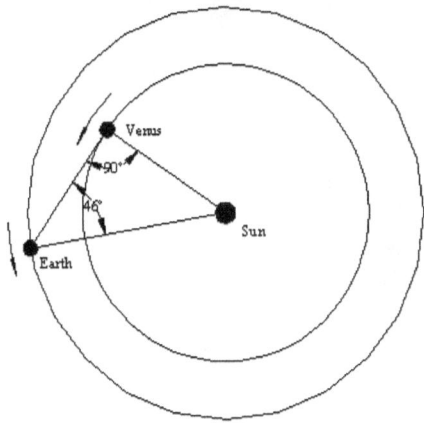

Fig. 3.4

The Earth, Sun, and Venus shown when the angle between the Sun and Venus is at a maximum.

Venus is moving around the sun at a faster rate than the Earth. As time proceeds following the fig. above, Venus will catch up with the Earth and the angle shown will be less than 46°. At a time before fig. 3-4, Venus was further away from the Earth and closer in the sky to the Sun. In this case the angle would have been also less than 46°. Thus the maximum angle from the Sun to Venus is ~46°. If the Earth-Sun distance is one *au*, what is the Venus-Sun distance?

K.   I can figure that one out; it is just a simple right triangle with an angle of 46° and a hypotenuse of 1 *au*. The Venus-Sun distance is 1 × sin 46 = 0.72 *au*

Mr. T.   Good! You could figure out the distance to Mercury in the same way but you would get a different answer each time you did it because the orbit is more elliptical. The other planets are more difficult because they are further from the Sun and the triangles used for computation become more complicated.

We will use Kepler's calculations and he has found a wonderful relationship that will make our task of finding how gravity changes easier. He has found that for all the planets, their period squared divided by their distance cubed is nearly a constant.

$$\frac{T^2}{R^3} = Constant \tag{3.4}$$

Some of his results are given in the following table.

Planet	Period (yr)	Ave.Dist. (au)	T^2/R^3 (yr^2/au^3)
Mercury	0.241	0.39	0.98
Venus	.615	0.72	1.01
Earth	1.00	1.00	1.00
Mars	1.88	1.52	1.01
Jupiter	11.8	5.20	0.99
Saturn	29.5	9.54	1.00

Table 3-1
Planet data

J.   That's fine but I have forgotten what we are trying to do. Help!

K.   We are trying to find out how gravity changes with different distances from the Sun. We have two relations for the acceleration of an object moving in a circle eq. (3-2) and eq. (3-3). We can put the information from Kepler's law, eq. (3-4) into either one and see what happens. I am going to use eq. (3-2) and substitute $R^3 \times Constant$ for T^2.

$$a = \frac{4\pi^2 R}{T^2} = \frac{4\pi^2 R}{Const \times R^3} \propto \frac{1}{R^2} \tag{3.5}$$

Mr. T.   We now see that the acceleration and therefore the force of gravity ($F = Ma$) on a planet is inversely proportional to the square of the distance to the Sun. The genius of Isaac Newton now comes into play. He argues that if there is a gravitational force on a planet from the Sun this force must also be proportional to the mass of the planet, and there must also be an equal but opposite force on the Sun by the planet. In fact the force on the sun should be proportional to the mass of the Sun. Because the two forces are equal and opposite, they must be proportional to both the masses. He further argues that because we have a moon that orbits us and Galileo showed us that Jupiter has many moons orbiting it, this idea of gravitational attraction should include everything that has mass. This is the universal law of gravity and can be expressed as:

$$F \propto \frac{M_1 M_2}{R^2} \qquad (3.6)$$

In words we can express it as: "*every body in the universe attracts every other body with a force that is proportional to the product of their masses and inversely proportional to the distance squared between them*".

J.   You say everything is attracted to everything else. Am I attracted to this building as I walk by? This building has a lot of mass.

Mr. T.   You are attracted to the building by a small force but you can't detect it.

Sometimes this law is written as an equation. You can always make a proportionality statement into an equation with a constant. In this case the constant is G or the universal gravitation constant.

$$F = G \frac{M_1 M_2}{R^2} \qquad (3.7)$$

Measuring its value turned out to be very difficult and was not done for more than 100 years after Newton put forth the law. The problem was that very strong electrical forces could not be eliminated and would cause errors with the experimental apparatus. Its measured value is:

$$G = 6.67 \times 10^{-11} \ Nm^2/kg^2$$

Gravitation is the first and the greatest of the four forces we shall study but it is also the weakest. It is the greatest because it acts on everything and permeates all of space. It can't be shielded, attenuated, or otherwise messed with. It is the weakest because the other forces we will deal with are much much *stronger* at close distances.

We have one more task to accomplish before you go back to the simulator.

Angular momentum

You know about regular momentum and its conservation law. We now need to consider the momentum of things that go round and round. We will try to keep it simple by starting with a single object. Our object is moving along in a straight line and we want to find its *angular momentum* (L) about some point. L is defined by the linear momentum times the perpendicular distance from its path to the point. $L = Mv\,R_\perp$ The diagram for this is shown below and I have chosen the motion of the object to be perpendicular to $R_\perp$.

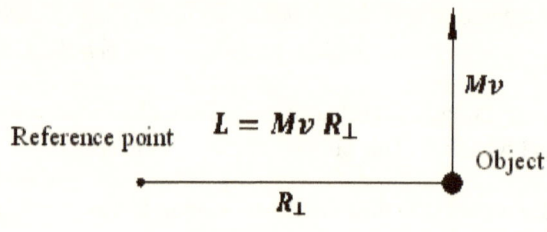

Fig. 3.5

Angular momentum of a single object
Note: the magnitude of L is given but the vector is not shown.

Now, we would like to show that if there is a force on the object directed toward the reference point (or center), the angular momentum will not change. A rigorous proof of this would require calculus and all of us don't have the skill required. Therefore, we will simplify the situation to a single impulse on the object directed toward the reference point or center. We will need to find an expression for L_1 (the angular momentum before the impulse) and L_2 (the angular momentum after the impulse).

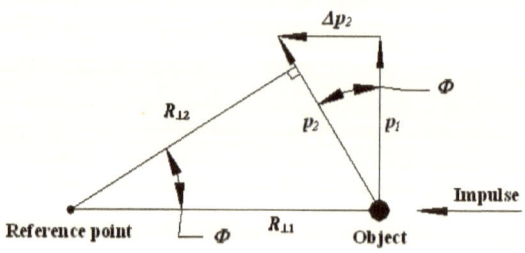

Fig. 3.6

Figure showing the momentum of an object before an impulse (p_1), and the momentum after the impulse (p_2) along with the perpendicular distances to the reference point ($R_{\perp 1}$ & $R_{\perp 2}$).

Why are the two angles, labeled Φ in the figure, the same?

K. The sides that describe them are mutually perpendicular; it is like rotating the one on the right by 90° to get the one on the left. What are we going to show with this diagram?

Mr. T. We are going to calculate the angular momentum before and after the impulse and show that it does not change. See if you can work through it.

K. From the vector triangle you can see that

$p_2 = p_1 / \cos \Phi$. The other triangle shows that:

$R_{\perp 2} = R_{\perp 1} \times \cos \Phi$. Now, what do I do?

Mr. T. Write the angular momentum before and after and use the relations you just gave us.

K. I will try. Here goes:

$$L_1 = L_2$$

$$p_1 \times R_{\perp 1} = p_2 \times R_{\perp 2}$$

$$p_1 \times R_{\perp 1} = (p_1 / \cos \Phi) \times (R_{\perp 1} \times \cos \Phi) = p_1 \times R_{\perp 1}$$

The cos terms cancel and the momentum before and after are the same.

J. But the planets are experiencing a continuous force from the Sun and not an impulse.

K. If the angular momentum does not change for a single impulse it won't change for thousands of little impulses all pointing toward the Sun or even a continuous force as long as it points toward the Sun.

J. I guess I can see that but I am missing the big picture. I sense that this "conservation of angular momentum" is important but I don't see why.

Mr. T. You will get to experience a little of its importance in the simulator. We will use it later in several places and I will give you some examples of where it is important, but for now, head for the simulator.

K.   I hope I get to do something this time. Last time Jill had all the fun.

Mr. T.   You both will be working this time and the lights will be on.

In the simulator

J.   We are back on the same frictionless plane and I am attached by a long rope.
Look, Kay is attached to the other end of this long (29 *m*) rope. She is going
slowly to the left and I am going to the right. The rope is tight. How fast are
we going?

K.   Take out your GPS device and read your speed. I will see if I also have one.

J.   I am going south at only 1 *m/s*.

K.   I am going north at 0.923 *m/s*.

J.   Why are we going at different speeds?

K.   You have a mass of 60 *kg* and mine is 65 *kg*, so it looks like the simulator
has set it up so that our momentums add to zero. 1 *m/s* × 60 *kg* = 0.923 *m/s*
× 65 *kg*. That means our center of mass will not be changing. We will only
rotate about it. That center will be closer to me (13.8 *m*) because I have more
mass (65 *kg*). Your distance to the center should be 15.2 *m*.

F.   Note: your momentums add to zero but your angular momentums add to a
greater number

   (1.74 × 10^3 *kg m^2/s*). Pull on the rope until you are closer.

J.   Let's pull until we are 4 meters apart; that's a good distance.

K.   OK! Here we go.

   It's getting harder and harder to pull the rope and we are moving faster and
faster.

J.   Wow! We are 4 m apart and really moving fast. Kay, read your GPS. I am
moving at 7.25 *m/s* and the force is too great. I have to let go.

K.   I didn't have to let go because when you let go the force went away. That
force was way too big. We can figure out a rough value for the force. The
acceleration is *v^2/ R* and the radius of your circle would be about half the

distance between us or 2 *m*. This gives an acceleration of $(7.25^2 / 2)$ or 26 *m/s²*. That is almost 3 times the acceleration due to gravity. The force on you was about three times your weight.

J.   I don't understand why we were going so fast.

K.   The forces we exerted on each other were always toward the center (center of mass). Therefore, angular momentum was conserved. We shortened the distance between us from 29 *m* to 4 *m* which is a factor of 7.25. Therefore, our speeds had to increase by the same factor.

J.   But that would mean our kinetic energy increased by a factor of 7.25^2. Where did all that energy come from?

K.   When we pulled we were exerting a force through a distance (shortening the rope). We created the kinetic energy. We were able to do this in a simulator but I don't think either of us could have pulled so hard in real life.

J.   Somehow, I can still hear you but I can't see you. We have been moving away from each other since you let go and now you are too far away for me to see. I think it's time to end this simulation.

Back in the classroom

Mr. T.   There are many things that involve angular momentum including: hurricanes, tornadoes, tops, stars, galaxies, figure skaters, atoms, nuclei, and many others. We will encounter a number of them later in the course. I have included a picture of a galaxy and I think you can see the evidence of rotational behavior.

K.   You have shown us that an object with a force directed toward some rotational center has an angular momentum that is conserved and does not change, but what does that have to do with a spinning top.

Mr. T.   When you add up all the angular moments of all the components of a spinning top, the sum will give you a total angular momentum for the top that must be conserved. This will give the top a very unusual dynamic behavior.

In order to understand this phenomenon, you need to recognize that the angular momentum *(L)* is a vector with both a magnitude and direction.

J.   How do you figure out the direction?

Mr. T.   For most situations it is along the spin axis and we use a right hand rule. Place your fingers around the axis with your fingers pointing in the direction of particle motion and your thumb shows the direction of the angular momentum.

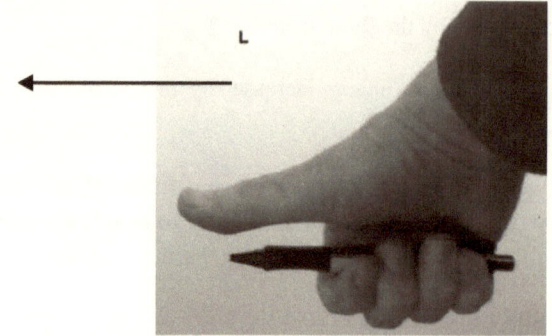

Fig 3.7

The right hand rule for angular momentum

Can you see how the rule applies to the top shown in fig 3.8?

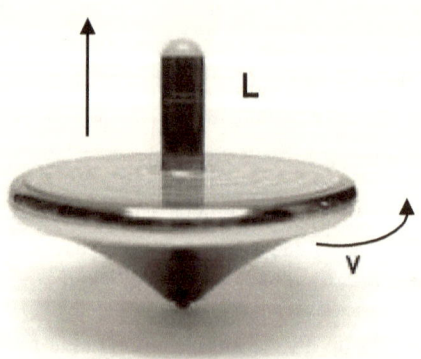

Fig 3.8

Angular momentum (L) and spin direction of a top

J.   I figured out the direction of the angular momentum vector but I don't see why the top doesn't fall over.

K.   I think I can answer that. The conservation of angular momentum means that the vector cannot change in either magnitude or direction unless acted upon by an outside force. If it tips, the angular momentum would change.

Mr. T.   Angular momentum is a complicated subject but I think you are getting the basic idea. Look at the picture of the spiral galaxy below. The total angular momentum will be the sum of all the moments of the stars. Can you figure out the direction of the angular momentum of this galaxy?

Fig 3.9

Spiral Galaxy NGC 1232

J.   It appears that the galaxy is rotating in a clockwise direction with the far away stars moving slower than the ones near the center. Therefore, if I am holding my hand correctly, the angular momentum vector is pointing into the page.

K.   It's just like it was when we were holding the rope; when we pulled closer, we went faster. When gravity pulls the stars in, they speed up.

Mr. T.   We have only one more topic before finishing this lesson. We need to understand something about gravitational potential energy and how it changes with distance from the Sun or any star.

J.   You told us there would not be so many calculations in this lesson as in the previous one. Now we are getting more and more.

Mr. T.   Sorry! I just can't stop myself and besides, you will need all this stuff. What do you remember about gravitational potential energy from last time?

K.   When we lifted the block the gravitational potential energy (U_g) was just the weight (Mg) times the height (h) we lifted it.

$$U_g = Mgh$$

Mr. T.   That was true at the surface of the Earth where g was a constant 9.8 m/s^2. Let's use a graph to illustrate what we are talking about.

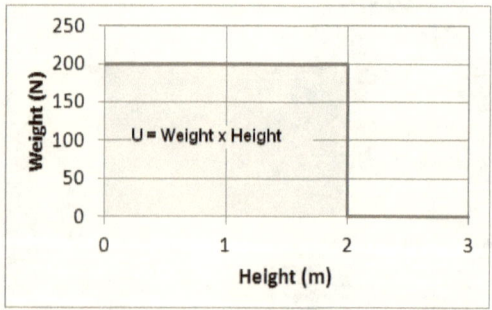

Fig 3.10

Graph showing U_g equal to an area

Unfortunately, when we move away from the Sun the weight (F_g) is changing inversley as the square of the distance.

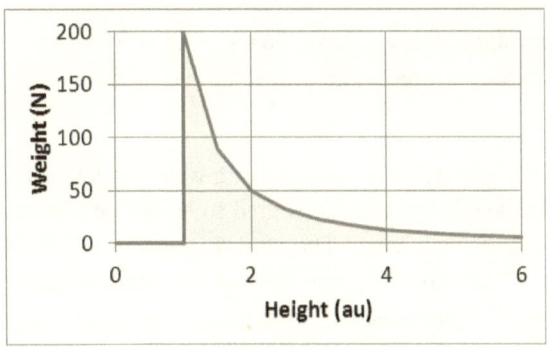

Fig 3.11

***Graph showing the weight of an object as a function of its distance from the
Sun in astronomical units (au)***

As we move the object away from the sun its gravitational potential energy
(U_g) will increase by an amount equal to the colored area of the graph. It is
usual that we consider the potential energy to be zero when the object is far
from the Sun and unable to detect any gravitational effects. This means that
the energy at a close distance must be negative.

J.   I don't understand what you mean by negative energy.

Mr. T.   It is negative <u>potential</u> energy. If you grab an object and move it far
away, you do work on it and give it energy. If we consider its potential energy
zero when it is far away, its energy when it is here must be negative.

K.   How do we find the area under a curve like fig 3-9?

Mr. T.   Look at the curve for a minute. You can see that the area under the far
parts of the curve gets smaller and smaller. Isaac Newton gives us the weight
of the object as:

$$Wt = G \; M_{sun} \; M_{object} / R^2$$

He is also able to show using calculus, which he invented, that the area under
the curve from a distance R to infinity is:

$$G \; M_{sun} \; M_{object} / R$$

Therefore, the gravitational potential energy is given by:

$$U_g = -G\,\frac{M_{sun}M_{object}}{R} \tag{3.8}$$

Note: The distance R is measured from the center of the Sun.

J.   How can he take an area all the way to infinity without it just getting bigger and bigger?

Mr. T.   It's like a frog trying to jump out of a well. He jumps half way out and holds on to the side but now he is tired so he can only halve the remaining distance with the next jump. The same thing happens with the next and the next jumps. Each time he jumps half the remaining distance. You can add up millions of his jumps and he always gets closer but he never gets out.

K.   Now that we have this expression for potential energy, what can we use it for?

Mr. T.   When we study the stars we will use it and we will use a similar expression when we look at atoms. Let's try a simple example now.

J.   Do we put our helmets on?

Mr. T.   This will just be a thought experiment.

J.   I can see it now, more calculations!

K.   Jill, don't whine.

Mr. T.   In the last lesson, while in the simulator, you threw a ball straight up and talked about its kinetic energy changing into potential energy as it went up. At the top, it had no kinetic energy and only potential energy.

Suppose you could throw the ball so hard it left the Earth and traveled straight out in space. For this situation we will neglect the Sun and the other planets. What have we learned that we can apply to this situation?

J.   I have been exposed to a lot; I have not learned it yet.

K.   Jill, be patient; you will get it.

The gravitational potential energy of the ball is given by:

$$U_g = -G\,\frac{M_E M_{ball}}{R}$$

Where M_E is the mass of the Earth.

And the kinetic energy by:

$$KE = M_{ball}\, v^2/\, 2$$

The total energy is the sum of the two:

$$E_{total} = -G\,\frac{M_E M_{ball}}{R} + M_{ball}\, v^2/\, 2$$

Now what do I do? What are we trying to do?

Mr. T.   Let's just look at your expression for the total energy and see what it can tell us about the ball. If Jill threw the ball very hard and the total energy was positive, what does that mean? What will the ball be doing far from Earth?

K.   As the ball goes up it will gain potential energy and lose kinetic energy. This will continue until the potential energy is zero (or near zero). If at this time the ball still has kinetic energy, it will move on in space forever. When the potential energy is zero, the gravitational force from the earth is small; there is no way to bring the ball back.

Mr. T.   If Jill threw the ball just fast enough for it to travel away and not return, we call that the escape velocity. We can calculate this escape velocity but we need the radius of the Earth (R_E) as well as G (the gravitational constant) and the mass of the Earth.

$$M_E = 6 \times 10^{24}\ kg$$
$$G = 6.67 \times 10^{-11}\ Nm^2/\,kg^2$$
$$R_E = 6.38 \times 10^6\ m$$

Jill, see if you can find the velocity.

J.   I will try if you help.

To just escape the total energy must be zero or:

$$U_g + KE = 0$$

$$M_{ball}\, v^2/\, 2 = G\,\frac{M_E M_{ball}}{R}$$

The mass of the ball cancels and I can solve for v^2.

$$v^2 = 2\,GM_E\,/\,R \quad \text{or}$$

$$v^2 = 2 \times 6.67 \times 10^{-11} \times 6 \times 10^{24} \,/\, 6.38 \times 10^6$$

$$v^2 = 1.25 \times 10^8 \quad \text{and}$$

$$v = 1.1 \times 10^4 \; m/s$$

Mr. T.   I was ready to help and you didn't need it. Good job!

J.   That's really fast. It must be the kind of velocity we need if we go to Mars. Is there anything other than a rocket that goes so fast?

Mr. T.   Remember in the last class when you were in the room with the bouncing argon atoms. A hydrogen molecule is much smaller and, therefore, would move faster. Let's see if it can move fast enough to escape the Earth. I will do the math for you. Last time we found that (eq. 2-9):

$$\frac{2\,KE}{3} = k_B\,T$$

Where KE was the kinetic energy of the molecule, k_B was Boltzmann's constant (1.38×10^{-23}), and T was the absolute temperature. By using the escape velocity we found, calculate the KE and determine the temperature required for the Hydrogen to escape. The hydrogen molecule has two atomic mass units (amu) and each amu is 1.67×10^{-27} kg. The KE is given by $M v^2 / 2$.

$$KE = 2 \times 1.67 \times 10^{-27} \times (1.1 \times 10^4)^2 \,/\, 2$$

$$KE = 2.0 \times 10^{-19}$$

The temperature is:

$$T = \frac{2\,KE}{3\,k_B} \quad \text{or}$$

$$T = \frac{2 \times 2.0 \times 10^{-19}}{3 \times 1.38 \times 10^{-23}} = 9.7 \times 10^3 \; {}^{\circ}K$$

K.   That's almost 10 thousand degrees. I don't think the hydrogen will escape.

Mr. T.   I don't either but in the early formation of the Earth, the temperature could have been even hotter and this may be the reason why we have so little hydrogen compared with Jupiter or Saturn.

I'm tired; I think we can we quit for today? See you next time.

Session 4

Electricity

If it weren't for electricity we'd all be watching television by candlelight.
George Gobel

If it weren't for electricity there would be no candlelight.
G Petersen

If it weren't for electricity there would be no candle.
Murgatroid Gundle P. Hagenheimer

In the classroom

Mr. T.   Today we begin the study of one of the strong forces in nature. Let's begin by finding out what you already know. I expect it is quite a lot.

J.   I know that electricity is all about charges. There are positive charges and negative ones. The like charges repel each other and the opposite charges attract. Charges can move through wires (electric current) but not through insulators.

Mr. T.   Not all charges can move through wires. What charges move through wires? Can you name the positive and negative charges that we know about?

K.   The charges moving in wires are electrons. The positive charges are protons and most of the time they can't move because they are fixed in the nuclei of atoms.

J.   A whole molecule can move if it's in the form of an ion in a water solution. An ion is a molecule or part of a molecule with one or more electrons removed or added. If an electron is added, it is a negative ion. If one is removed, we have a positive ion.

K.   In a plasma, the electrons get stripped off and currents can flow by moving both electrons and positive ions. This happens in a fluorescent light bulb.

Plasmas, liquids, solids, and gases, are sometimes called the four states of matter.

Mr. T.   Do you know the units of electricity?

J.   I know some; there are amps, volts, and watts, but I am not sure of their definitions.

Mr. T.   I am sure you know much more but we need to start with very basic concepts and build some models.

J.   Mr. Tweed, what do mean by models?

Mr. T.   If we say a gas is like a bunch of round molecules bouncing around, hitting the walls and each other, that is our model for a gas.   Using mathematics and the model, you can make predictions about the behavior of the real world.

K.   But molecules are not just balls. They have shapes and can rotate and vibrate.

Mr. T.   You are correct, but we will always start with simplest model and see where it takes us.

Electric Fields

Mr. T.   I am sure you know something about fields but my guess is that your knowledge has not been formalized.

K.   What do you mean by formalized?

Mr. T.   The field has a very definite mathematical meaning to the scientist. Your understanding of a field may not be very specific. If your understanding was formalized, the field concept would have a very definite meaning.

K.   I get it. Let's do it.

Mr. T.   OK!   Let's consider a small positively electrically charged object. Normal objects have a lot of charge but they are neutral because there is as much negative charge (electrons) as positive charge (protons). If we have a positively charged object, some of the electrons will have been removed, leaving excess positive protons. The amount of charge on the object (q_1) will be the excess protons.

We will visualize a field around this object as a series of field lines that radiate out in every direction from the object. These lines will show us the direction of force on a small positive charge placed in the field.

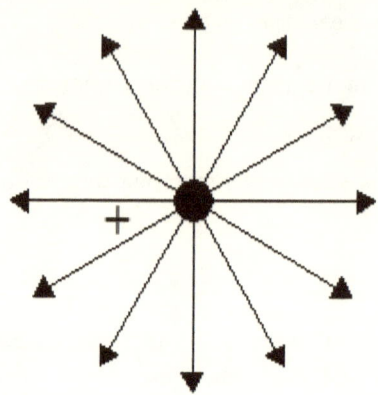

Fig 4-1
Electric field lines from a positive charge.

J.   The picture tells us the direction but what about the magnitude?

Mr. T.   It will even tell us something about the magnitude. If the charge on the object is q_1, think of there being q_1 field lines. Now consider that the density of lines (number of lines per area) will tell us how strong the field is. The area we choose needs to be perpendicular to the field lines. If we draw a sphere around the object with the center at the object, the density of field lines intersecting the sphere will be constant and as the radius of the sphere gets larger, the field will get smaller. How will the density of field lines change?

K.   The area of a sphere is $4\pi r^2$ and the number of lines is constant, and all must pass through a sphere, so the density of lines (number per area) will be inversely proportional to the radius squared. The field lines will also be perpendicular to the surface of the sphere. This indicates that the field strength (E) will vary as $1/r^2$ and because the number of lines depends on the charge of the object (q_1), E will also be proportional to q_1.

$$E \propto \frac{q_1}{r^2}$$

J. We have a proportional statement and, therefore, we can turn it into an equation. All we have to do is change the proportional symbol to an equal sign and add a constant of proportionality.

$$E = k\,\frac{q_1}{r^2}$$

K. We have a field equation. How do we find the force on some charged object in the field?

Mr. T. The force will depend on the charge on the object (q_2) in addition to the field (E).

$$F = E \times q_2 = k_C\,\frac{q_1 q_2}{r^2} \qquad (4\text{-}1)$$

K. Why did you put the "C" subscript after the constant?

Mr. T. This constant is called Coulomb's constant after the French physicist **Charles Coulomb**. The unit of charge or Coulomb (**C**) is also named after him.

$$k_C = 8.988 \times 10^9\,\frac{Nm^2}{C^2}$$

Equation 4-1 is known as Coulomb's law. One Coulomb of charge is equivalent to a very large number of elementary charges. It was defined before it was known that charge comes in small packages equal to the charge on an electron or proton. At that time, charge was thought to be like a continuous fluid. We use e to indicate an elementary charge.

$$6.25 \times 10^{18}\,e = 1\,C$$

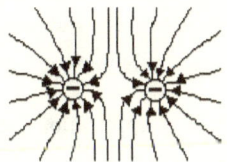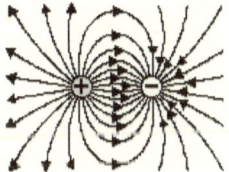

Fig. 4-2
Fields of other charge arrangements

Mr. T.   Notice the fields from these charge arrangements. It is interesting to consider how the field appears from a great distance. Jill, what does the field of the two negative charges look like from a distance?

J.   If you go far enough away, you will not be able to tell if there are one or two charged objects. The field lines point toward the charges because a positive charge will be attracted.

Mr. T.   Correct! Kay, what can you say about the plus-minus pair?

K.   If you go to a great distance, all the field lines that start at the positive charge will end at the negative one and you will see nothing. The world is full of positive and negative charges which cancel each other out, and from a distance we are unable to detect any electric field.

Mr. T.   Good! Now let's look at another charge distribution. There are two plates. The top plate has a negative charge and the bottom one has an equal positive charge.

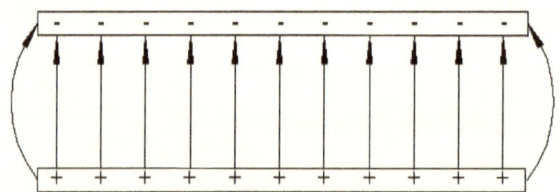

Fig. 4-3
Field from two charged plates

Mr. T.   Jill, tell me about this field.

J.   The field lines are parallel. Therefore the field density is the same near the top plate as it is near the bottom one. This means the field strength is constant from the top to the bottom. There are no field lines above the top plate or below the bottom one. The field is almost entirely between the plates.

Mr. T.   This arrangement is useful for a number of applications. Suppose I drill a small hole straight through both plates, put it in a vacuum, and boil off some electrons near the top hole.

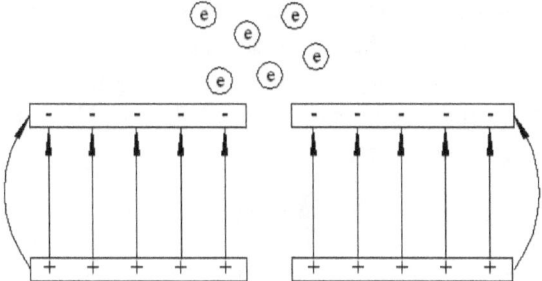

Fig. 4-4
Cross section showing hole and electrons

K.   Will the electrons be repelled by the negative plate? Will they move into the hole? If they can get into the hole, what happens?

Mr. T.   Suppose I let you look at the situation as seen by the electron.

"Hey electron, I am a negative plate and if you get close I will send you up," said the negative plate.

"I don't think so, the positive plate is telling me to go down, so I am not going to do anything," said the electron.

"My child, enter the hole and you will know which way to go." said the Genie in the hole.

"OK", said the electron. "Whee! Both plates are telling me to go down. That was fun and now I am going really fast, but where are the plates," said the electron.

"You have passed through both holes and are now free; the plates are way above you," said the Genie in the hole.

J.   I understand some of it. The electrons will not be repelled by the negative plate while they are outside because there are no field lines except between the plates. Because you boiled them, they must be moving and some may enter the hole.

As soon as they enter the hole they will feel the electric field and be accelerated downward. The force on them will be -*Ee* where *E* is the electric field strength and —*e* is the charge on an electron. If they are accelerated downward and pass through the bottom hole, they will be free and have no force on them. If this process is continued, there will be a beam of electrons coming from the bottom hole.

However, I don't understand how you boil electrons.

Mr. T.   In a metal, electrons can move around; that's what makes a metal a conductor of electricity. If you heat the metal and jiggle the electrons to and fro, some will move so fast that they escape and leave the surface. It is as if they were boiled off.

If some electrons were accelerated by the field between the plates, they were given kinetic energy (***KE***). What would you need to know in order to figure out how much kinetic energy?

K.   We have an expression for the force (-***Ee***) so we just need the distance which in this case is the separation between the plates (-***h***).

$$KE = Eeh = eV \qquad (4.2)$$

Therefore, we need to know the charge on an electron, the distance between plates, and the electric field strength.

J.   Why did you use —***h*** for the distance?

K.   Because the electron was displaced in a downward direction. Both the force and the displacement were in a downward (-) direction, therefore, the product is positive.

Mr. T.   Correct! It is time to introduce a new quantity. Try to follow me. The energy per charge in your kinetic energy expression is ***Eh***. We have a term for this (***V***) and it is called the voltage. Batteries are one good source of voltage for the laboratory. The chemical action of the battery will move electrons to the negative terminal and remove them from the positive terminal. If an electrical load such as a light bulb is connected to the battery, electrons will flow (current) from the negative terminal through the light bulb to the positive terminal. If we have a 6 Volt battery, each Coulomb of charge that flows through the light bulb will release 6 Joules of energy. Most of this energy will be in the form of heat but some will be in the form of light.

In our situation, if we connect a battery to our plates, current will flow until both plates are charged up to the battery's electrical potential energy (voltage). The voltage from one plate to the other is:

$$V = Eh \qquad (4\text{-}3)$$

J.   This equation tells me that the voltage is dependent on the electric field and the distance between plates but when you connect the battery, the voltage is just dependent on the battery. I am not sure I understand.

Mr. T.   You understand perfectly. Another way to read equation 4-3 is to say the electric field depends on the voltage and inversely on the distance between the plates.

J.   Let me see what that means. If the voltage is fixed by the battery or voltage source, and the distance between plates is increased, the electric field will go down.

K.   This must mean that the charge on the plates is less when the plates are moved apart. That's true because the number of field lines depends on the amount of charge and if the density of field lines is low, the charge producing these lines must also be low.

Mr. T.   There are many things about these plates that we could investigate but I think it is now time for another simulation.

In the simulator

J.   I see that we are in a very large room with a ceiling about 3 *m* high, but I don't see anything else. Fritz, are you there?

F.   As long as the power is on, I am always here. Welcome back! The room you are in appears to be *3 m* high but it is actually only *5 mm* or $5x10^3$ *m* high.

K.   So we are really very small.

F.   You are not only small but the only thing in the room you can interact with is the large knob that controls the voltage between the floor and the ceiling. The meter in front of you will show you what the voltage is between the ceiling and the floor. When the ceiling is positive, the voltmeter reads positive and the electric field lines point down.

J.   I still don't see anything other than the meter and knob.

F.   I'm sorry; I have to turn the bright light on.

K.   What do you mean you are sorry? Computers don't forget things and they don't have feelings.

F.   My programmers must have wanted me to wait until you asked before turning them on and I am programmed to be polite. Do you see anything now?

J.   I see hundreds of extremely small particles shining in the bright light with the dark walls in the background. They are falling slowly toward the floor.

F.   They are very uniform plastic spheres about 1×10^{-6} *m* in diameter. In this room they appear to be slightly larger than ½ *mm*. They are sold to calibrate microscopes and have many biological uses.

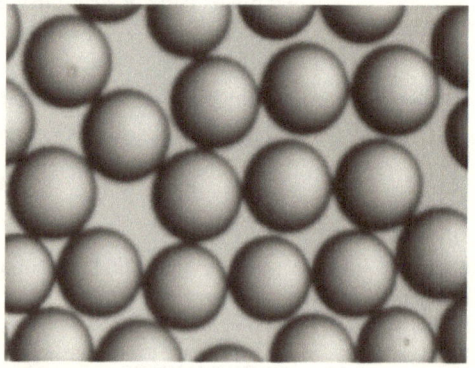

Fig. 4-5
Electron microscope picture of uniform latex spheres (diam. = 1 micron, $M = 5.5x10^{-16}$ kg). Magnification is approximately 12000.

J.   0.5 *mm* is small and they are falling right through me as though I wasn't even here.

F.   You are like a hologram. I told you that you would not interact with objects in this room.

The object of this simulation is to demonstrate that charge comes in small packages and is not a continuous quantity. We are using small spheres because they are so small that a change of just one elementary charge will change their behavior. Spheres even smaller than we are using have the problem that they would fall too slowly and we would have trouble seeing them. In the original experiment, Robert Millikan[1] used oil drops which come in a continuous variety of sizes. The plastic spheres represent a huge advantage over oil drops because they all have the same size and mass.

K. Why are the spheres moving downward with a constant velocity? Why are they not accelerating at - 9.8 m/s^2?

F. They are moving at terminal velocity. If you sky dive, you will accelerate until you reach about 55 m/s and then fall at this constant rate. Air resistance keeps you from falling faster.

J. Obviously everything does not have the same terminal velocity. What determines the rate that things fall?

F. The density and temperature of the air, the mass, and the object's surface area. When a sky diver jumps from a very high altitude, the air is thin and he will fall very fast. When his parachute opens, his surface area is radically increased and he slows down. A filled balloon falls slowly, but a bowling ball with about the same size will accelerate for a long time.

There is an interesting ratio which is helpful in understanding many physical phenomena, the surface to volume ratio (*Area/Volume*). If we compare objects with the same spherical shape and material, this ratio will vary as the radius is changed. The area and volume of a sphere are given by:

$$Area = 4\pi r^2$$
$$Volume = \frac{4\pi r^3}{3}$$

Therefore,
$$\frac{Area}{Volume} = \frac{3}{r}$$

The *Area/Volume* ratio gets very large when the radius of the object is very small.

The small spheres, you have been using, have an enormous surface area compared to their mass. Therefore, they fall very slowly.

K. It all seems simple enough. Why are you expending such an effort to explain it?

F. Many people find the concept non-intuitive; you need to think about it in order for it to make much sense. You will need it in later sessions.

Now let's see what happens when you apply the electric field. Turn the knob and see what happens.

J. They were all falling but now most are moving up with different speeds.

F. Pick one that is moving fast and by adjusting the position of the knob see if you can stop it.

J. Look at that! I stopped one and now there are many that are sitting there motionless in space. However, none of them were moving very fast, how can I tell if I really stopped them or are they just moving very slowly?

F. When you get them stopped, I will change the clock speed so you can make fine adjustments and Kay can record the voltage.

J. Tell me again what I am doing when I move the knob.

K. I can do that. You are adjusting the voltage until the electric force on the sphere is equal to its weight. You are stopping many spheres because they all have the same charge.   Your first reading is -33.6 V.

F. Jill, increase the voltage, locate the next fastest sphere and stop it. If too many of your spheres are getting too close to the floor or ceiling, you can bring them back by adjusting the voltage and even reversing it.

J. It is a similar situation just like before; many have stopped and not just the one I was looking at.

K. Your second reading is -42 V.

F. Repeat as many times as you can with different spheres.

K. We found five different groups of spheres and the voltage readings are as follows:

-33.6   -42   -56   -84   &   -168 V

Fritz, how can we analyze these results?

F. Before you analyze the results, please remember what the object of this simulation is.

K. We are trying to show that charge comes in discrete quantities.

F. There is no requirement to do any calculations in order to accomplish that. All you have to do is look carefully at the data.

J. I can see that. We obtained five discrete values for the voltage to stop the spheres. If the charge on the little plastic objects could be any value, we would have found many more than five results. I think we did it.

F. Now look at the groups and answer the question: Which group has the greatest charge? Which group has the least charge and why.

J. The group that I stopped with 33.6 *V* had the most charge because it required the least voltage to stop. It is possible that the group requiring 168 *V* had only one excess charge. If that's true, maybe the situation is:

Voltage	Charge (*n*)	*Vn*
-33.6	5	-168
-42	4	-168
-56	3	-168
-84	2	-168
-168	1	-168

K. Why did you multiply the voltage times the number of charges?

J. In order to stop the sphere, the total force on it must be zero. The weight is *Mg* and the electrical force is *Ene* where *E* is the electrical field strength, *e* is the elementary charge unit, and *n* is the number of these charge units on a sphere. The field strength (*E*) is *V/h*. This gives us:

$$Mg + \frac{Vne}{h} = 0 \qquad\qquad (4\text{-}4)$$

All the terms in this equation are constant except for *V* and *n*. Therefore, the product of *V* and *n* should be constant for the equation to work.

K.   Because you have figured out the values for **n**, we can solve equation 4-4 for the charge of an electron (**e**).

$$e = \frac{-Mgh}{nV} = \frac{-5.5\times10^{-16}\times(-9.8\times5\times10^{-3})}{1\times(-168)}$$

$$e = -1.6 \times 10^{-19} \; Coulombs$$

J.   That looks good, but you didn't use my other results. You only used **n** =1.

K.   They will all give the same result; try it.

J.   That's ok, I believe you.

F.   I hope you realize that in a real experiment there will be variations and you would not get exactly the same result for different particles.

This simulation is over. See you next time.

In the classroom

Mr. T.   I hope you now have some feeling for the fact that charge comes in discrete amounts. We have a name for things that come only in definite quantities; they are **quantized**.

There are many things to learn about electricity and I have given you some pretty difficult material to start with. Would you like to try something easier?

J.   Mr. Tweed, could you give us something really easy?

Mr. T.   I will try. Let's look at a simple circuit. The symbol for a battery is shown on the left. The symbol for a light emitting diode (LED) is on the right and has little arrows showing the emission of light. The solid triangle shows the direction of electrical current. The lines represent wires and the switch symbol is in the *off* position. If it were in the *on* position the line would touch both circles (contacts).

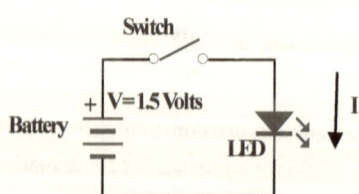

Fig. 4-6
Simple circuit diagram

In a circuit, electrons flow from a source (battery) through some wires and a load (LED). They must flow back to the battery after passing through the LED and switch. Kay, can you tell me why?

K. Electrons are negative and if they were to flow only away, the battery would soon become charged to a positive voltage.

Mr. Tweed, I have a question. The electrons must leave the negative terminal, which is the bottom one in the diagram, and flow to the positive one on the top. Why is the LED arrow pointing down when the electrons must move up to get to the positive terminal?

Mr. T   That's a very good question. When scientists first began to study electricity, they didn't know whether positive or negative charges were moving. They adopted a sign convention assuming positive charges were moving. There have been attempts in the past to point the arrows in the direction of electron flow but these attempts have always been rejected. In fact, in a solution or plasma both positive and negative charges are flowing in opposite directions.

The flow of either positive or negative charges is called an electric current. The unit of electrical current is the **Ampere** (*A*), and one Ampere is equal to one Coulomb per second (*C/s*). The symbol for electrical current is *I*.

J. I can tell Mr. Tweed; you have given us some new symbols and that means we are going to see some more math.

Mr. T.   You are very perceptive. I did promise to keep it simple, and I will.

What could you change to increase the current (*I*)?

J. The only thing I can see that could be changed is the battery. I would use a battery with a greater Voltage.

K. I imagine some diodes would conduct electricity better than others, so you could change to a different diode.

Mr. T.   That's true. The parameter that tells us how well something will conduct electricity is its resistance *(R)*. A scientist by the name of **Ohm** worked out the simple relationship:

$$V = R\,I \qquad\qquad (4.5)$$

This relationship is known as Ohm's law.

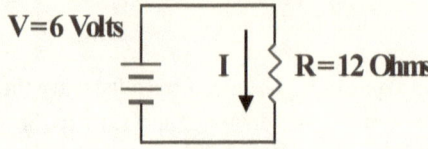

Fig 4.7
Simple circuit with resistance and battery
The resistance symbol is the jagged line.

Jill, can you tell me how much current is flowing in the circuit above?

J.   I can do that in my head. The current is *V/R or* 0.5 *Amps*.

K.   The rating on the battery is 6 *V* (*Joules/Coulomb*). Where is the energy going?

J.   I can answer that. All the energy goes into heating the resistor.

K.   But I know the battery will get hot or at least warm

Mr. T   You are correct. The battery has some small internal resistance which is not shown the circuit diagram. Assuming no heat is generated in the battery, how much heat will be added to the resistor.

K.   It will depend on how long you leave the circuit connected. In one second 0.5 *Coulombs* flow through the resistor and each one will generate 6 *Joules* of heat. Therefore, the heat will be 3 *Joules/sec*.

Mr. T.   The rate of generating heat or energy is called power and the units are *Joules/sec* or *Watts*.

$$Power = V\,I \qquad\qquad (4.6)$$

Two other expressions can be derived by using Ohm's law. Equation 4.4 can be solved for V or I and the results substituted into equation 4.6.

$$Power = VI = \frac{V^2}{R} = I^2R \qquad (4.7)$$

M r. T. Look at the following circuit. The circle with the sine wave represents an alternating current (AC) generator usually located at the power company. When alternating current is used, the current flows back and forth and not in one direction as it does with a battery. The circuit breaker will interrupt (stop) the current flow if it exceeds 15 *Amps*. Most circuit breakers in homes are rated at 15 *Amps*. Three electrical loads are plugged in to this circuit in a house. Calculate the current, resistance, and power for each one. What is the total current flowing through the circuit breaker?

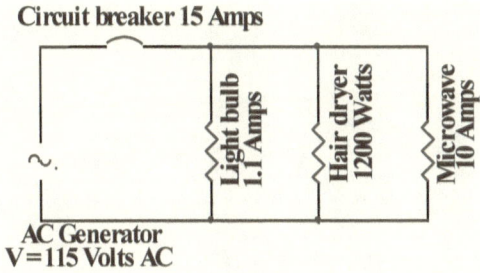

Fig 4.8

Single circuit in house with three appliances connected (plugged in).

J. I will do the light bulb. The power is easy. *P=VI=115x1.1=127Watts.* The current and voltage are given so: *R=V/I=115/1.1=105 Ohms.*

K. The power of the hair dryer is given so:
I=P/V=1200/115=10.4 Amps.
And *R=V/I=115/10.4=11.1Ohms.*

J. The microwave is the same problem as the light bulb.
P=VI=115x1.1=127Watts.
And *R=V/I=115/10=11.5 Ohms.*

K. The total current through the breaker is the sum of all three. *I$_T$ =* *1.1+10.4+10=21.5 Amps.*

It is clear that there is a problem. This is more current than the circuit breaker will allow. The breaker will trip and all three appliances will go off.

Mr. T.   Almost all appliances that produce heat require high currents. Therefore, the wiring for these circuits use bigger wire and the circuit breakers are designed for more current.

J.   Are we done for today?

Mr. T.   I think so. See you next time.

Session 5

Magnetism

I happen to have discovered a direct relation between magnetism and light, also electricity and light, and the field it opens is so large and I think rich.
— <u>Michael Faraday</u>

In the simulator

J.   This session has just started and we are in the simulator already. Fritz, what are we doing in here?

F.   Today you are here to learn about magnetism. You will see very small things and your clocks have been adjusted so that you can observe them moving very fast.

Kate, there are two boxes in front of you. Each one contains a positive charge. Put one charged ball in each hand and hold them.

K.   I have them and they are repelling each other and pulling my hands apart.

F.   Now let go of them and observe their motion carefully.

K.   They are accelerating away from each other. At first they had a pretty high acceleration but now it's less. However, they will be out of sight soon.

F.   I have retrieved them and will now accelerate them to a very high velocity using an apparatus similar to the one you used in the last session. You will see them coming together from your right side. Are you ready?

J.   Do we have to make any measurements or do any calculations?

F.   No, you only have to pay attention to how the particles move apart compared to what they did when Kate let them go.

K. Here they come. They are together and moving to the left fairly quickly. As they move their paths are getting further and further apart. They are gone. Can we see that again?

F. OK. Here they come.

J. They are clearly separating from each other but not nearly as fast as they did when Kay let go of them. Why not?

F. This is a simple demonstration of what you will be studying in this session, magnetism. These charges are being subjected to magnetic forces that are trying to hold them together.

J. This is nothing like anything I have learned about magnetism before. What is going on?

F. I will give a quick explanation and after you have studied the material in the rest of this session, it should make more sense.

Each moving charge produces a magnetic field. The other charge moves through this magnetic field and experiences a force of attraction which tries to counteract the electric force of repulsion. The magnetic force tries to pull the two charges together. Their paths are becoming more parallel.

K. I assume this magnetic force depends on the velocity. Could the speed be high enough to pull the charges together?

F. No. That could never happen. However, there is a velocity that would just counteract the electric repulsive force and the charges would not move apart.

K. What velocity would be required to do that?

F. If the charges were both moving at the speed of light, the magnetic attractive forces would exactly cancel the repulsive electric ones.

J. But Einstein said nothing but light can move at the speed of light.

F. That's correct, so I don't think it would be a good idea to program the simulator to simulate any motion greater than or even equal to the speed of light.

J. So, what should we have learned in this simulation?

F.   This simulation was just preparation for learning a number of things about magnetism. You will learn that magnetic fields (**B**) are generated by moving charges and these fields act on moving charges.

In the classroom

Mr. T.   Magnetism has a number of very interesting features. It is difficult to know where to begin. I think we should start by looking at the force on a single moving charge.

J.   You have told us that magnetic fields act on moving charges. Therefore, I would expect the force to be proportional to the velocity of the moving charge (*v*).

K.   I would also expect that it depends on the charge of the moving object.

Mr. T.   Correct and that is about all that is involved other than figuring out the direction of the force.

$$\vec{F} = q\,\vec{v} \times \vec{B} \tag{5.1}$$

This is a vector equation and the × symbol indicates something called a cross product. Only components of **B** and *v* that are perpendicular to each other are considered and the force is at right angles to both of them. The usual way for us to deal with situation is to take the component of **B** which is perpendicular to *v* and multiply it by *v*.

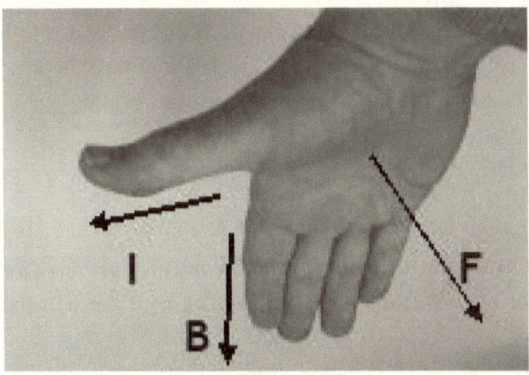

Fig. 5.1
The right hand rule

This diagram shows the directional relationship between the magnetic field (**B**), the current (**I**), and the force (**F**).

The vector along the thumb is labeled **I** but it also works for **v** when you have a single moving charged object.

J.   Does it matter if the charge is negative or positive?

Mr. T.   If the charge is negative, you must reverse the direction of your thumb.

K.   Why is there no constant of proportionality in equation 5.1?

Mr. T.   Both **B** and **q** have been defined in such a way that the constant is one.

J.   If we are studying a moving charge, why does the diagram have a vector labeled **I** and not **v**.

Mr. T.   A current is just a bunch of moving charges which are usually but not always confined to move in an electrical conductor (wire). We take up currents right after your next simulation.

In the simulator

F.   Hi again. I am all ready for you.

J.   What are we doing today?

F.   You are going to play ball with single electrons. You will be made very small and your clock will be speeded up so that you observe very fast motions. Actually, you will not throw bare electrons because you would not be able to see them. We will attach electrons to balls that have negligible mass. You can only find balls like this in a simulator.

K.   How big is our space and how much faster is our clock?

F.   Your room is **20 cm (0.2 m)** on a side and your clock is 10^8 times as fast as normal. There is a vertical magnetic field of **1x10^{-3} Tesla** which can be turned on or off. A **Tesla** is the unit of magnetic field. In fundamental units it is $\dfrac{kg}{Coulomb\ sec}$.

The magnetic field is now off. Kate, take one of the balls and throw it straight across the room. How much time did it take to reach the opposite wall?

K.   It took slightly more than one second to travel 15 *cm* (0.15 *m*) which means that the velocity is about **0.15 *m/s*** .

F.   Don't forget that your clock has been speeded up by **10^8**. The actual speed is **1.5×10^7 *m/s***. I will now turn on the magnetic field and Kate will throw the ball again.

K.   Wow, it went in a big curve to the left and crashed into the left wall.

F.   Try standing near the middle of the south wall and throw the ball in the east direction.

K.   OK. Now the ball moves in a complete circle. It started east, turned north, then west, and finally east again before I caught it.

F.   Did it change speed as it went around?

K.   I don't think it did. It seemed to move with constant speed with a diameter slightly less than the size of the room. It took about **3.5 *s*** to go around. It moves sort of like a planetary orbit but there is no object at the center. Fritz you can probably tell me what the radius of curvature was.

F.   It was **8.5 *cm***. When a planet moves in a circular orbit, what direction is the force on it?

J.   Toward the sun or the center of motion and it is always at a right angle to the direction of motion. We learned that in session 3.

F.   I told you the magnetic field was in a vertical direction but I did not say if it was up or down. Use the right hand rule and figure out if the field is up or down. Remember, Kate threw a negative charge, so the thumb should point opposite to the direction of motion.

J.   I am first going to guess that the field is down so I will point my fingers down. My thumb should point west since Kate threw the ball east. My palm is now pointed south which means I guessed wrong. The ball curved toward the north. Therefore, the magnetic field points up.

K.   I get it. If I point my fingers up and follow the path of the ball, my palm always points toward the center of motion.

F.   Kate, try throwing the ball slower and don't catch it when it comes around.

K.   OK.   It continually moves in a smaller circle but the time to go around remains about **3.5 s**. If I don't catch it, it just keeps moving in a circle.

Fritz, the faster I throw it the larger the radius of curvature. If we measure the radius, can we determine the speed?

F.   Let's work out the math.   You just learned that the magnetic force on a moving charge (Eq. 5.1) is:

$$F = Bqv$$

This assumes everything is at right angles. In session 3 (equation 3.3), you learned that the acceleration of an object moving in a circle was given by:

$$a = \frac{v^2}{R}$$

This means the force is:

$$F = \frac{m\,v^2}{R}$$

We now have two expressions for the force. If we set them equal to each other, we get:

$$Bqv = \frac{m\,v^2}{R}$$

One of the velocity terms cancels out and we have: $mv = p = BqR$ (5.3) This tells us that if we measure R we can calculate the momentum. If we know the momentum and the mass, we can get the velocity.

Mr. Tweed has some more examples for you. See you next time.

In the classroom

Mr. T.   I can show you the path of a real electron from a bubble chamber in CERN.

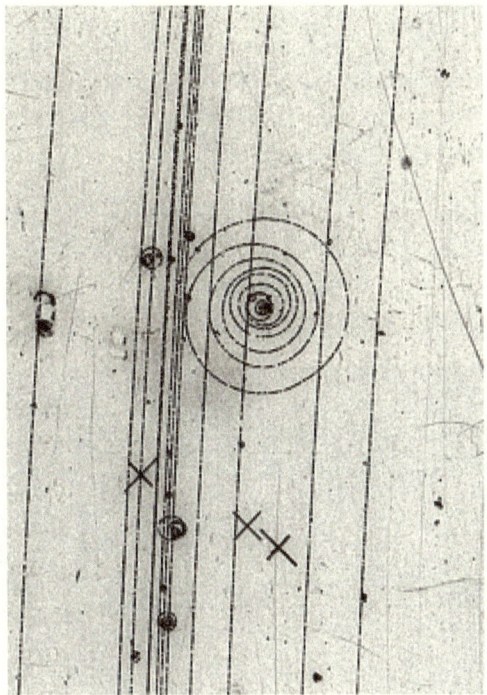

Fig. 5.2

Picture showing a spiraling electron as seen in the CERN 2-metre hydrogen bubble chamber.

J.   What is a bubble chamber?

Mr. T.   It is a chamber containing liquid Hydrogen at **14° K**. The pressure is suddenly reduced so that the Hydrogen is ready to boil just before sending in a stream of high energy charged particles. As the particles move through the Hydrogen they ionize the molecules causing very small bubbles to form along the path of the particle. The high energy particles enter from the bottom of the picture and move in almost straight lines.

Once in a while an electron on a Hydrogen atom is hit and sent flying. That's what you see in the bubble chamber picture. See if you can figure out the direction of the magnetic field. If we knew the magnitude of the magnetic field and the scale of the picture, we could determine the radius of curvature and, therefore, the momentum of the electron (Eq. 5.3).

K.   The electron's radius gets smaller as it moves which means it is losing momentum and energy. Why?

Mr. T.   There are two reasons. Some of the energy is lost ionizing the Hydrogen molecules and some is lost because the electron is radiating electromagnetic energy. We will come back to this later.

Now let's look at an example that is of great importance to our lives. Charged particles are constantly being expelled from the surface of the sun. This flux of particles is called the solar wind and explains part of why the tails of comets are always directed away from the sun.

Fig. 5.3

Picture of a comet showing the tail being pushed away from the sun by the solar wind

Frequently there are storms on the sun's surface that expel huge amounts of material at very high velocities (much greater than the escape velocity of the sun). This greatly increases the solar wind.

K.   Why are the particles charged?

Mr. T.   The sun is so hot that atoms don't exist except at a distance from the surface. The material in the sun consists of bare atomic nuclei and electrons which are both charged.

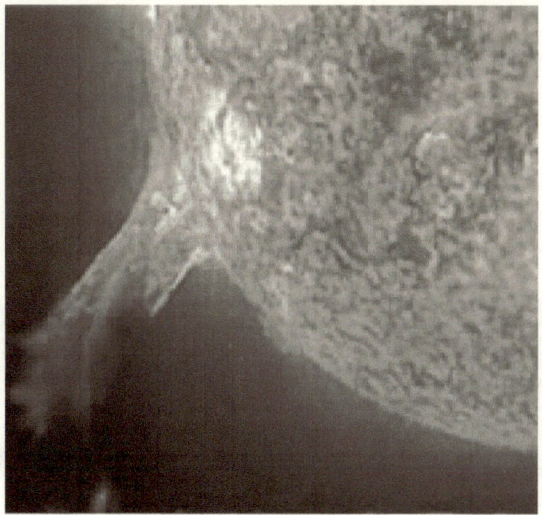

Fig. 5.4
NASA picture of a solar flare (storm)

Don't damage your eyes trying to look for such storms. They are best viewed by using special instruments which utilize parts of the electromagnetic spectrum outside our visible range. The picture in figure 5.4 was taken using an x-ray telescope.

Fig. 5.5
Another NASA picture of a solar storm

J.   Do these solar flares really affect us?

Mr. T.   Many of the charged particles travel all the way to earth and are capable not only of penetrating our atmosphere but can pass directly through us. When this happens they can ionize material in our cells and cause damage.

J.   That sounds terrible; what can we do?

Mr. T.   We can't do anything. It has been going on as long as the earth has been here. However, we have some protection. The earth has a magnetic field which is effective in deflecting these particles.

Fig. 5.6
Artist's picture of the earth's magnetic field

The magnetic field lines point toward the magnetic north pole which is why a compass points north.

Any charged particle coming from the sun will encounter a force as it enters the earth's magnetic field. Assume a positive charge enters the field of figure 5.6 from the right. Use the right hand rule and figure out the direction of the force on the particle.

K.   I got it; my fingers (*B*) are pointing up; my thumb (*v*) is pointing to the left and my palm (F) points into the page. This means positive charges would be deflected toward the east. My thumb would point opposite to the direction of negative charges and, therefore, they would go west.

J.   In the summer the North Pole is tilted closer to the sun. What happens then?

Mr. T.   Remember the charged particle must have a velocity component perpendicular to the magnetic field. This means the magnetic shielding is not as good near the pole; in fact the particles may follow the field lines down to the earth. In the north during a solar flare, many charged particles will enter the atmosphere ionizing the molecules of Nitrogen and Oxygen which will make them glow and cause the Aurora Borealis (northern lights).

Note: The up-to-date description of the Aurora involves knowledge of the collisions between the wind and the charged particles in the high altitude belts of the Earth's upper atmosphere. It is more complicated than my simple explanation. However, I hope you get the main idea.

Fig. 5.7
The Aurora Borealis as seen in northern Norway

J.   Do all the planets have magnetic fields?

Mr. T.   Mars and Venus have no magnetic field now but we have evidence that Mars had one in the past.

J.   If we send people to Mars, will they be in danger without the shielding of a magnetic field?

Mr. T.   They will have to stay in a shielded space much of the time especially if there is a solar storm. They will also need protection during their flight in a space ship.

J.   I don't want to go.

Mr. T.   Neither do I!

It is probably a good time to discuss where magnetic fields originate. We can start by examining some phenomenon that you are probably familiar with. You have played with small permanent magnets. What can you tell me about them?

K.   Each magnet has a north pole on one end and a south pole on the other. If you can find a way to allow the magnet to move freely, the north end of the magnet will point in a northern direction.

Also, the north pole of one magnet will attract the south pole of another and repel its north pole. This means the north pole of the earth is actually the south pole of a magnet. The north pole of a permanent magnet is attracted to it.

Mr. T.   Good! Jill, can you add anything?

J.   Permanent magnets are made from iron and must be electrically polarized before they will work. If you heat them, they will lose their magnetism.

Mr. T.   Actually there is a whole class of materials, with what we call ferromagnetic properties that can be made into permanent magnets. Nickel and Cobalt are good examples.

Permanent magnetization is not required for these materials to exhibit magnetic effects. A permanent magnet can induce a temporary field in a piece of raw iron and cause it to become a magnet. If small oblong pieces of iron (iron filings) are placed in a magnetic field, very interesting things will happen. Look at the following picture and tell me what you see.

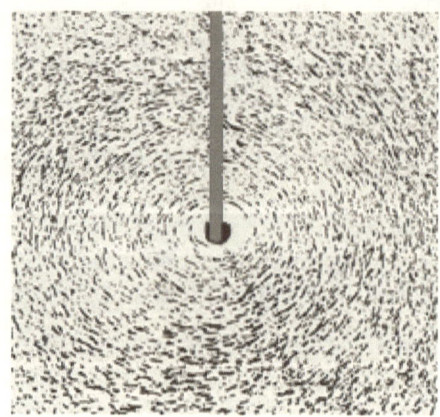

Fig. 5.8
A current carrying wire is passed vertically through a paper that has had iron filings sprinkled on it.

J.   The filings will point in the direction of the field and the north end of each will attract the south end of the iron filing in front of it. This shows us that the magnetic field forms in circles around the wire. The picture also shows us that the field strength is stronger close to the wire because the iron filings tend not to align when they are away. However, we don't know in which direction it is going (CW or CCW).

Mr. T.   You have been able to see every bit of information available from Fig. 5.8. I have another picture which shows us the direction but we still have to figure out how the field varies with distance from the wire.

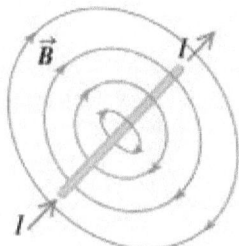

Fig. 5.9

Direction of the magnetic field around a current carrying wire

Mr. T.   We can use another right hand rule to describe this direction.

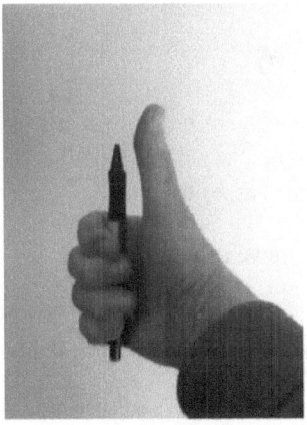

Fig. 5.10

Right hand rule for magnetic field direction from a current carrying wire

The pen represents the wire with the thumb in the direction of the current (**I**) and the fingers represent the field (**B**) around the wire.

The relationship between the magnetic field strength (**B**) and the current (**I**) in the wire is simple. $B \propto I$. We need to do an experiment to determine how **B** changes with distance from a wire. Fritz awaits you for the experiment.

In the simulator

F. Good to see you back.

J. You are just a computer. I didn't know you could see.

F. I can see very well when I am connected to a camera. Unfortunately I can only interpret what I see when I have been programmed to do so. Mr. Tweed has programmed me to recognize you.

K. What are we going to do today?

F. Something very similar to what we did at the beginning of this session. We started with positive charges moving parallel to each other and saw that their motion caused a force that opposed the electric force of repulsion.

Today our charges will move in wires and we will not see the repulsive forces.

J. Why don't we see the repulsive force?

F. Kay, can you answer her?

K. I think so. The wires contain both positive and negative charges in equal numbers. The electric field at a distance will cancel and we will see only the magnetic effects of the charges that move.

F. Jill, what charges can move in a wire?

J. Only the negative electrons can move; the positive nuclei are fixed in the crystal structure of the metal.

F. Fig. 5.11 is a very simple representation of our experiment. The wires are one meter long (**L=1m**) and separated by one **cm** (**R**=10^{-2} **m**). A current of one **Ampere** is flowing to the right (electrons are moving to the left) in each wire. At this time we and the diagram will only consider the magnetic field generated by the left wire. Jill, can you describe using the right hand rule the direction of the field where it comes in contact with the wire on the right.

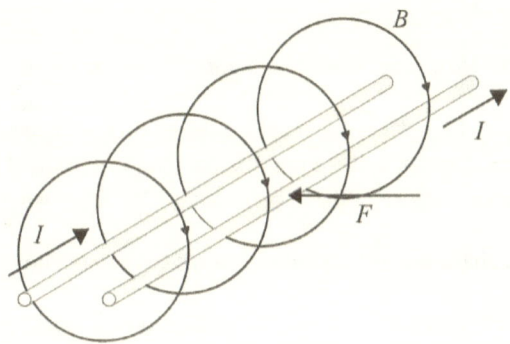

Fig. 5.11
Diagram for magnetic field measurement experiment

J.   If I place my thumb along the left wire in the direction of the current, the magnetic field is directed in a clock-wise fashion. It is pointed down where it comes in contact with the wire on the right.

F.   Now use the other right hand rule of Fig. 5.1 to show the direction of the force on the moving charges in the wire on the right.

J.   Using my fingers for the magnetic field (pointing down on the right wire) and my thumb for the current (in the right wire), my palm shows the force to be toward the left or the same as in the diagram.

K.   I understand the direction but we found an expression for the force, $F = Bqv$, and we don't know the charge (q) or the velocity (v).

F.   That's correct, but we will measure the current (I) and the length of the wires (L). We will look at all the charge (Q) available to move in the length (L). The current should be: $I = Q / \Delta t$. Δt is the time for all the charge (Q) to move through the length (L) of the wire. Therefore,

$$Q = I \Delta t \text{ and } v = L / \Delta t.$$

Kay, substitute **Q** for *q* and the expression for *v* into your expression for force and see what you get.

$$F = Bqv = B(I\Delta t)x(L/\Delta t) = BIL \qquad (5.2)$$

K. I think we have to remember that **B** and **I** must be at right angles for this formula to work.

F. That's right. Now what should we figure out before we do the experiment?

J. We know **I** and **L** but you have not explained how we can measure the force shown in the diagram and we don't know the magnitude of **B**. I don't even know what the purpose of the experiment is. What are we trying to do?

F. Kay, do you have any suggestions as to what we might do?

K. If you can show us how to measure the force on the length **L** of wire, we can use equation 5.2 to determine the magnitude of **B**. However we still don't know the relationship between **B** and the distance from the wire (**R**). If we know that, we might be able to determine the expression for **B** around a wire. How do we measure the force?

F. I will place sensor blocks against the wires which are interfaced to the computer. These blocks will detect the force of the one meter wire pushing on them and also keep the wires from moving together.

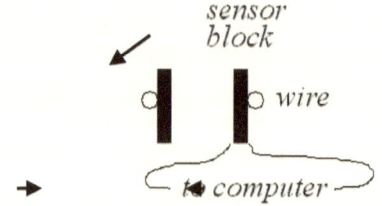

Fig. 5.12
Sensor blocks for measuring the force on length L of wire and maintaining proper separation

Jill will switch on the current and Kay can read the force on the computer screen. You can then change the distance between the wires and repeat the measurement to see how the field (**B**) changes with distance.

J.   I am not sure I understand but here goes. The current of one Amp is flowing in each of the wires.

K.   The computer tells me there is a force of

$2x10^{-5}$ N on the one **meter** length of wire. Because both the length L and current I are equal to 1, B is equal to $2x10^{-5}$ **Tesla.**

Now let's increase the separation from 1 **cm** to 5 **cm**.

J.   I moved the sensor block for a separation of 5 **cm**. and the current is on.

K.   The force now reads $4x10^{-6}$ N so the field B is $4x10^{-6}$ **Tesla**. The force dropped by a factor of 5 when we increased the distance by a factor of 5. We have an inverse relationship.

$$B \propto 1/R \qquad\qquad (5.4)$$

Also I am going to guess that the field B is directly proportional to the current I. Fritz, is anything else involved, or can we just calculate a constant of proportionality and declare victory?

F.   If you don't change the medium between the wires, you can declare victory. The constant you calculate will work for a vacuum and most materials.

K.   The relationship should be:

$$B = C\frac{I}{R}$$

If values for I, B, and R from the first experiment are used in this expression, we get:

$$2x10^{-5} = \frac{1}{10^{-2}}C$$

Therefore, the magnetic field of a long straight wire in a vacuum is given by:

$$B = 2x10^{-7}\frac{I}{R} \qquad\qquad (5.5)$$

J.   We declared victory but what did we do that was so important?

K. 1. We showed that magnetic fields come from moving charges or currents.

2. We showed that magnetic fields act on moving charges or currents.

3. We calculated the force on a charge moving in a magnetic field.

4. We calculated the force on a current carrying wire in a magnetic field.

5. Finally, we calculated the magnitude of the magnetic field generated by a straight wire carrying a current.

J. You left out the most interesting part. We learned how some things with vector properties act on each other at right angles and not in the direction you would expect. The idea that we need right hand rules is just plain weird and very non-intuitive. Fritz, are there other things that act like that?

F. There are many strange things in physics and some of them are weirder than you can even imagine. That's one of the things that make the subject so interesting.

K. We did address some important ideas but there are a lot of topics on magnetism that we did not even mention. Why does the earth have a magnetic field? Why can you magnetize iron and not aluminum? My mother had some Magnetic Resonance Imaging pictures taken in the hospital. How does that work? What about transformers? Fritz, are we going to continue studying magnetism?

F. Some topics will come up later but we will not have a complete session devoted to the topic. You will have to use the internet or a book to investigate some of these topics.

J. Are we done today?

F. We have only one more quick simulation.

K. Suddenly it is very dark; I can barely see anything. All I see is a large bare room.

F. The simulator is again making you seem very small and your clock is speeded up so that you can see fast moving objects. Jill you see that small charged ball in front of you? Pick it up and throw it as hard as you can at the wall. Tell me what you see.

J. Ok, I am throwing it.

Wow, there was a flash of light when it hit the wall. What caused that?

F. When the electrical charge hit the wall there was a large acceleration. When charges are accelerated they produce electromagnetic waves. Light is an electromagnetic wave and waves are the topic for the next session.

Session 6

Waves

The tendency of modern physics is to resolve the whole material universe into waves, and nothing but waves. These waves are of two kinds: bottled-up waves, which we call matter, and un-bottled waves, which we call radiation or light. If annihilation of matter occurs, the process is merely that of un-bottling imprisoned wave-energy and setting it free to travel through space. These concepts reduce the whole universe to a world of light, potential or existent, so that the whole story of its creation can be told with perfect accuracy and completeness in the six words: 'God said, Let there be light'. — Sir James Jeans

In the classroom

Mr. T.   I am glad to see you back again. Today we will begin studying the properties of waves. This investigation will lead us into areas that are familiar and easy to understand and also into phenomena that are downright weird and very non-intuitive. Let's start by examining what you already know. Jill, do you want to start?

J.   I know there are water waves, sound waves, light waves, radio waves, ultraviolet waves, infrared waves, microwaves, and possibly more.

Mr. T.   That's pretty good, you named all the really important ones. Did you know that several of the waves you named are really just one type?

J.   Which ones are the same type?

Mr. T.   Light, radio, ultraviolet, infrared, and microwaves are all electromagnetic waves. However, they have different wavelengths.

J.   Mr. Tweed, I think I understand "wavelength" but it might be a good idea if you went over the concept.

Mr. T.   For many waves the wave pattern is repeated many times as it travels. The distance (in *m*) between patterns is called a wavelength and we usually use the Greek letter λ for this variable. As the wave passes a particular point the wave pattern will repeat itself after a time (*T*) which we call the **period** of the wave. Figure 6.1 shows how the electric field strength (vertical axis in arbitrary units) changes with distance along the direction of propagation. This picture captures the wave in an instant of time.

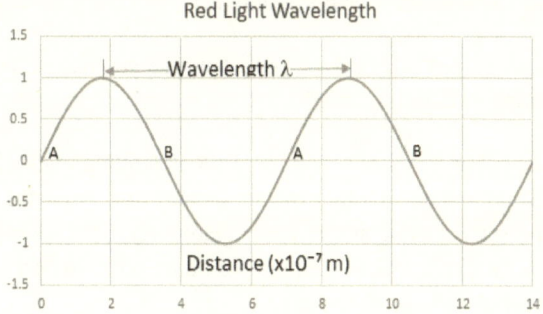

Fig. 6.1

Crests and troughs of the electric field of red light. Wavelength is 7×10^{-7} m

Figure 6.2 shows the electric field strength of the same red light, only at a specific point along the direction of propagation. This diagram shows how the field changes with time.

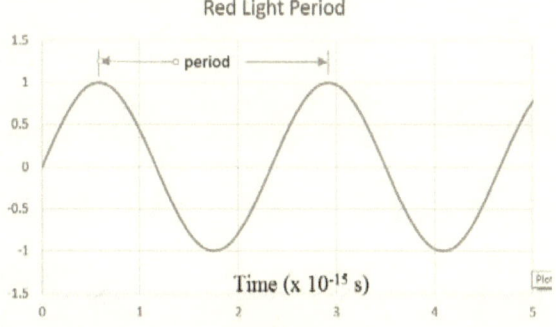

Fig. 6.2

The electric field of red light as a function of time.
$T = 2.333 \times 10^{-15}$ s

J.   The two pictures look the same. It is confusing. How do we make sense out of it?

Mr. T   It is not difficult but many people get confused by these two pictures. Kay, can you figure them out?

K.   In figure 6.1 you see something like a photo of the wave frozen in time. In figure 6.2 we see how one point on the wave changes with time. However, it is not clear what the starting position is. The electric field is zero at zero time which could be positions *A* or *B* in figure 6.1.

Mr. T.   If we assume it is one of the *A* positions, can you determine whether the wave is moving to the right or left?

K.   I will try. Looking at figure 6.1, if the wave is moving to the right (positive *x* direction) and we take another picture a short time later, the electric field will have a negative value at that time. However, after a short time figure 6.2 indicates the electric field strength would be positive. This is wrong, so the wave must be moving to the left.

J.   Are you sure? I didn't follow all of that.

K.   Try this. Pick up the wave in figure 6.1 and move it a short distance (0.1λ) to the left. The field value at position *A* can be determined by where the line (wave) crosses the y axis (above position *A*). It has a positive value. Now look at figure 6.2. At time zero it has a value of zero. A short time later ($0.1T$) when the picture is taken, the value is positive, which is what we predicted.

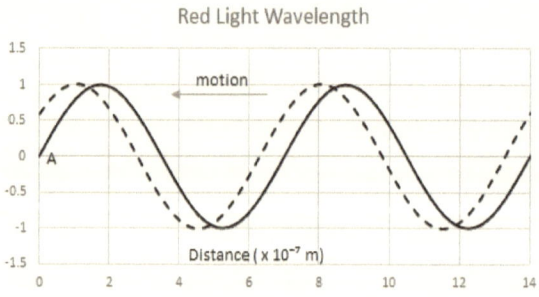

Fig. 6.3

Wave shown after T/10 seconds delay

J.   In the graphs the electric field goes positive and negative. What does that mean?

Mr. T.   In this case when the value is positive the electric field is pointing up and when negative it points down. It is polarized in an up-down direction. An electromagnetic wave is a *transverse* wave which means the field direction is at right angles to the propagation direction. Therefore, the wave could be polarized in any direction such as into and out of the paper. However, the polarization must be perpendicular to the wave motion.

A sound wave is a *longitudinal* wave which means the molecules move in the same or opposite direction to the wave motion. It can't be polarized.

J.   What about water waves?

Mr. T   Water waves are very complicated and the water molecules move back and forth as well as up and down. Their motion also changes with depth.

We are now in a position to write down a very important relationship between the wavelength (λ) and the period (T). We know how far the wave travels in a period so we can write the expression for the velocity (v). Kay, give it a try.

K.   This is easy. The velocity is just the distance divided by the time or:

$$v = \lambda / T \qquad\qquad (6.1)$$

Mr. T.   If we recognize that the frequency ($f0$) equals $1/T$, we can rewrite Eq. 6.1 as:

$$v = \lambda f \qquad\qquad (6.2)$$

J.   It's not clear to me why the frequency is the reciprocal of the period.

K.   It's easy if you use some simple numbers. If the frequency is two waves per second, the period must be half a second which is the reciprocal of two. I guess I should say the frequency is 2 **Hertz** (Hz) because Hz is the accepted unit for frequency.

Mr. T.   Jill, look at figures 6.1 and 6.2. They will give you both the period and the wavelength for red light. See if you can calculate the velocity of light.

J.   That's easy. The wavelength (λ) is 7×10^{-7} *m* and the period (T) is 2.333×10^{-15} *s*. The velocity can be found from equation 6.1:
$$v = \lambda / T = 7 \times 10^{-7}/2.333 \times 10^{-15} = 3 \times 10^8 \ m/s$$
We probably should use *c* for the velocity symbol because we are talking about light.

Mr. Tweed, at the end of the last session we threw a charged ball at the wall in the simulator and it made a flash of light. Can you explain more about how we make light waves?

Mr. T   I will let Fritz try in the simulator.

In the simulator

F.   Hi! I understand that I am supposed to show you how to generate an electromagnetic wave. Electromagnetic waves all travel at the speed of light so we should speed up your simulator clock by a factor of 10^8. Jill, take hold of the negatively charged ball again. There is a field all around the ball pointing toward the ball.

J.   I don't see any field lines.

F.   You can't see the electric field lines but, fortunately, I have some magic electric field line sprinkles. I will blow them into the air around you.

J.   Wow, there are lots of lines radiating out from the ball.

F.   I am now going to remove the sprinkles from all the lines but one, the one pointing horizontally directly away from you.

K.   Strictly speaking, isn't the line pointing toward Jill and not away because the charge is negative.

F.   You are correct; unfortunately the sprinkles don't show the direction of the field.

Jill, move the charged ball sharply downward about half a meter and tell me what you see.

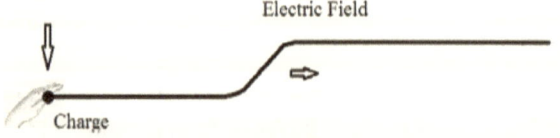

Fig. 6.4
Wave in the electric field line of the simulator traveling away from Jill after she moves charge downward.

J.   I moved the charge down but not quite 0.5 *m*. The field line had a bend in it that moved away fairly rapidly. After a short time it was positioned horizontally from the new charge position. I can see a wave in the electric field but why is it called an electromagnetic wave?

F.   Remember from the last session what happened when you moved a charge.

K.   We produced a magnetic field.

F.   What did Jill just do?

K.   She moved a charge so she must have generated a magnetic field in the process. I assume it would travel out with the wave in the electric field line.

F.   That's correct; every time there is a wave in the electric field, there is an accompanying magnetic wave. They go together; it is an electromagnetic wave. Note that both the electric and magnetic components are perpendicular to the propagation direction. These are transverse waves.

It may be a little easier to visualize if we think of a wire which contains both positive and negative charges. Right now let's look at some of the field lines from just the negative charges. The positive charges will have field lines in the opposite direction that cancel out the lines from the negative charges but we will neglect them now.

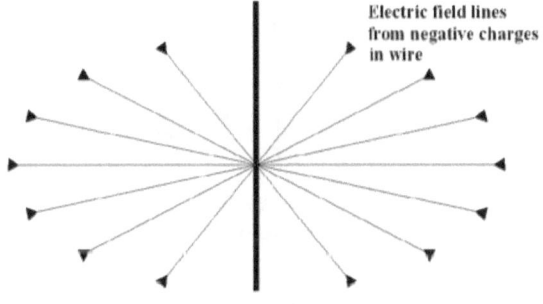

Fig. 6.5

Some of the field lines surrounding a wire from the negative charges

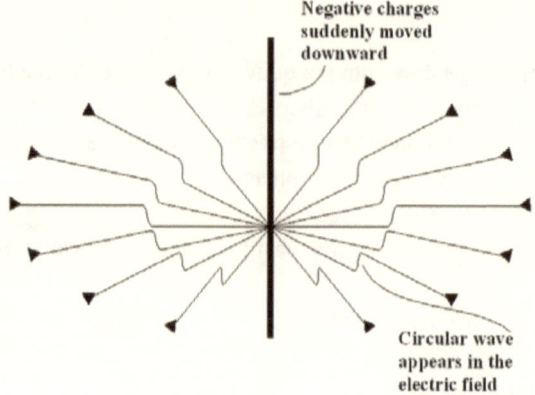

Fig. 6.6

The field lines shortly after the negative charges are displaced downward

F.   If we now also consider the field from the positive charges, they will cancel out the negative field lines and the only thing remaining will be the wave portion. Jill, use the right hand rule to verify that the magnetic field is counter clockwise when viewed from above.

J.   The negative charges were moved down so my thumb must point up and my fingers are pointing counter clockwise (ccw).

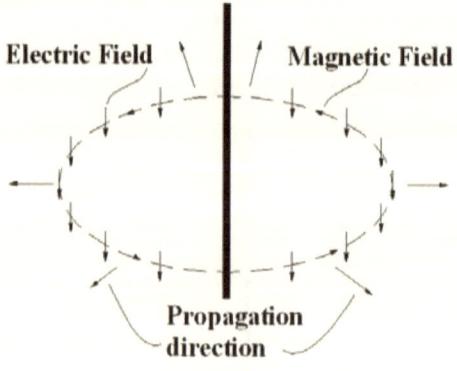

Fig.6.7

Diagram showing both the electric field wave (short arrows pointing down) and the circular magnetic wave (dotted line) shortly after the negative charges (electrons) were displaced downward in a wire.

F. This is how low frequency radio waves are generated. The radio transmitting antenna is a long vertical wire and currents are caused to move up and down the wire at the proper frequency. The waves then move out like ripples on a pond from the antenna.

J. That's very interesting but my question before we entered the simulator was about light and not radio waves.

K. I think they are the same thing; the only difference is the frequency and wavelength.

F. Right! The electromagnetic spectrum covers a huge range of wavelengths and frequencies. Here is a table showing the rough features of the spectrum.

THE ELECTROMAGNETIC SPECTRUM

	λ (meters)	f (Hz)
Radio	>1	$< 3x10^8$
Microwave	< 1	$> 3x10^8$
Infrared	$\sim 10^{-5}$	$\sim 10^{13}$
Visible	$< 7x10^{-7}$	$> 4.3x10^{14}$
	$> 4x10^{-7}$	$< 7.5x10^{14}$
Ultraviolet	$\sim 10^{-8}$	$\sim 10^{16}$
X-Ray	$\sim 10000000^{-10}$	$\sim 10^{18}$
γ-Ray	$\sim 10^{-12}$	$\sim 10^{20}$

Table 6.1

Wavelengths and frequencies of the electromagnetic spectrum.

J. That's all very good but my question was about making light and I don't think that your explanation works very well for a light bulb or even a laser.

F. Have patience; we will get there eventually.   The important thing to remember is that electrons and positive nuclei are everywhere and the electric fields from these charges are everywhere. If any charge is jiggled, there will be a corresponding wiggle in the electric field which we call radiation. If the frequency of this radiation turns out to be in the visible spectrum, our eyes can detect it and we call it light. When current heats the filament of a light bulb, all the atoms and electrons are rapidly jiggling and sending light waves at many frequencies. We see this as white light which contains all the colors. Actually, most of the radiation from an incandescent light bulb is in the form of invisible infrared waves. We can't see it but we can feel it.

Even our warm bodies have molecules in motion which give off infrared radiation.   Some animals can image this radiation and see at night. Unfortunately, we can only do this with special night vision goggles.

What am I doing? I am supposed to operate the simulator and Mr. Tweed is supposed to be the teacher. Perhaps you should go back to class now.

J.   Thank you! We appreciate you teaching us. See you next time.

In the classroom

Mr. T. Hi!   I understand that you know how electromagnetic waves are produced.

J.   Fritz was able to show us a few things but the field of electromagnetic radiation is big and I don't think I understand very much. I know you want to move on but can you give us another concrete example and say a little more about creating electromagnetic waves.

Mr. T.   I can say a few thing about X-Rays and maybe that will help. An X-Ray tube consists of a glass globe which has been evacuated, a cathode designed to give off electrons, and a target anode (Tungsten, Copper, or possibly another metal) which collects the electrons after they have been accelerated by a very large voltage ($\sim 70 \times 10^3$ V). The electrons will give off some radiation as they are accelerated toward the anode but the X-Rays will be emitted when the electrons experience a large de-acceleration as they hit the anode.

The same thing happens in a computer crt (cathode ray tube). When the electrons hit the face of the computer screen, they not only cause the phosphors to light up, but also emit X-Rays. These X-Rays have low energy but it is probably a bad idea to sit too close and too long in front of a TV or computer screen. The newer computer monitors and TV screens don't have this problem.

Jill, you are correct; this is a large area of interest and we will return to it many times before we are finished.

J.   Thank you.

Mr. T.   Before we go too far there are a couple of simple properties of waves that require attention. Waves reflect when the medium they travel in changes and they can superimpose on each other. This means a wave traveling north can proceed unimpeded even though another wave is traveling west. They can move through each other and there is no reflection.

We can learn some interesting things about wave reflection by making some waves, here in the classroom, with a rope.

J.   This is a change. Are we going to do experiments in the classroom instead of the simulator?

Mr. T.   I don't think Fritz will mind if it is just this once. Kay, take hold of the rope attached to the wall, pull it fairly tight, and try to send a short pulse down it.

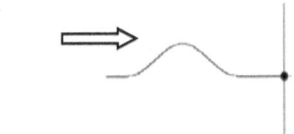

Fig. 6.8
Short wave pulse in a rope

Fig. 6.9
Inverted reflected wave from wall

K.   The wave traveled quickly toward the wall and then was reflected back. However, when it came back it was inverted.

Mr. T.   Let's try it again only now we will insert a slide that will allow the rope to move freely when encountering the wall.

Fig. 6.10
Wave traveling on a rope with a free end toward the wall

Fig. 6.11
Short pulse in rope as it hits the wall

Fig. 6.12
Non-inverted reflected wave

K.   When the wave was free the reflection had the same polarity as the incoming wave. Mr. Tweed, can you explain why that happened?

Mr. T.   I am going to let you think about that for a while. Now we will try to examine how waves interfere with each other or even themselves.

What do suppose happens when two waves with the same wavelength hit each other head on.

K.   You already told us; they will pass through each other and go on their way.

Mr. T.   That's true but what happens at the instant when they are moving through each other.

K. I suppose that in an instant when one wave says the field must be up and the other says down, the field will be confused and won't do anything. However, later the situation will have changed.

J.   I get the idea but I don't think the use of *confused* as a metaphor works very well. I think the instructions from both waves must just add together.

Mr. T.   They do add. We can examine the situation by taking snapshots of two waves with the same frequency moving in opposite directions. We will start when the waves are exactly on top of each other.

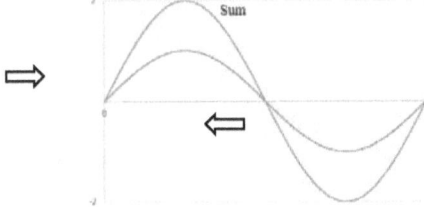

Fig. 6.13

Two waves at an instant when their phase angles are identical. The sum is just twice the value of one of them and has a value of zero at each end and in the middle.

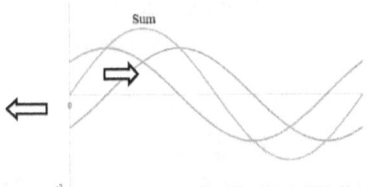

Fig. 6.14

In this picture the wave moving to the left has moved by λ/8 and the wave moving to the right by the same distance. After these waves are added, the sum is zero at both ends and in the middle.

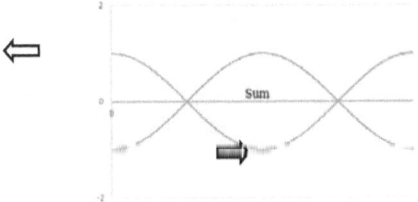

Fig. 6.15

Here both waves have moved by λ/4 and are, therefore, half a wavelength out of phase. Note that the sum is not only zero at both ends but zero everywhere. The two waves are cancelling each other.

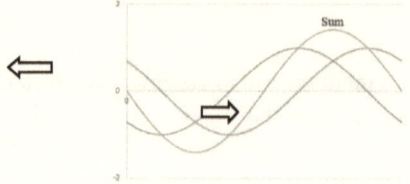

Fig. 6.16

A movement of an additional λ/8 yields this picture. Note the sum is still zero at the ends and middle.

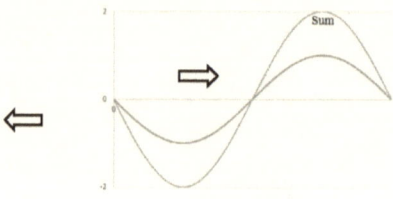

Fig. 6.17

The total movement of each wave is now λ/2, which puts them back in phase with each other. The zero points for the sum of the two waves are still at the beginning, end, and middle. The two waves add to an amplitude of two but is negative (down) on the left and positive on the right.

Mr. T.   I hope you can see that the two waves produce (add up to) a wave that does not move right or left but only goes up and down. Such waves are called standing waves. The places where they do not move are called nodes and the places of maximum motion are anti-nodes. One way standing waves are produced is to vibrate a string. The vibration sends waves traveling along the string in both directions and reflections occur at both ends.

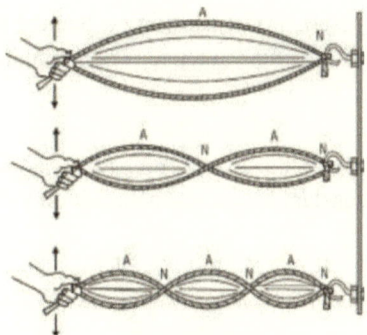

Fig. 6.18

Standing waves on a rope. In this case the person generating the wave must vibrate the rope at a resonant frequency. This diagram shows the fundamental, second, and third harmonic resonances. The anti-nodes are labeled A and the nodes N.

Mr. T. Kay, if the person vibrating the rope 0.8 *Hz* at its fundamental frequency and the length of the rope is 2.4 *m*, what is the wave velocity of the rope?

K. I know I need both the wavelength (λ) and the frequency to calculate the speed of the wave. I am not sure if the wavelength is 2.4 *m* or twice that number.

J. I can answer that; if you were to take a snapshot of the first rope in figure 6.18, you would see only an up or a down portion of the wave. You need both a positive and a negative portion to have a full wavelength. Therefore, 2.4 *m* represents only a half wavelength.

K. I got it. I will use equation 6.2 for the calculation:
$$v = \lambda f = 4.8 \times 0.8 = 3.84 \ m/s$$

Mr. T. Jill, can you tell me the frequency of the third harmonic shown in figure 6.18?

J. That's really easy Mr. Tweed; it's the third harmonic so it is three times the fundamental or 2.4 *Hz*.

Mr. T. That's probably very close to the correct number but it might not be exact. What assumptions did you use to come up with your answer?

J. I didn't make any assumptions; I just know that the third harmonic is three times the frequency of the fundamental. The harmonics are all just multiples of the fundamental.

Mr. T. Unfortunately, that's not always true. Look again at equation 6.2. What quantity has to be constant for your generalization to be true?

J. According to the equation if you divide the wavelength by three you must multiply the frequency by three in order to obtain the same velocity.

Mr. T. That is exactly correct but suppose the velocity for the harmonic was not the same as the fundamental.

J. Mr. Tweed, are you trying to make this hard?

Mr. T. No, but I don't want to make it too simplistic. The harmonics of most string instruments have close to the correct harmonic frequencies. However, the large strings that produce the low frequencies of the piano don't act like perfect strings and have harmonics that are a little off. A church bell has a particular clang to it because the harmonics are not even close to multiples of the fundamental.

Water waves are also complicated. The wave speed is dependent on the frequency, the water depth, and the wave amplitude. This helps us understand why waves break near the beach.

K. Mr. Tweed, I know that musical sounds contain harmonics and these harmonics are mainly responsible for the characteristic sound of each instrument. What can you tell us about that?

Mr. T. We could have a whole years' course on the physics of music. I will say a few things and if the subject interests you, you can do some reading.

A violin string is interesting and produces many harmonics. The position of the bow can control the relative strength of the harmonics. If the string is bowed close to the bridge (end), the harmonics will be strong. If it bowed closer to the middle, the fundamental will dominate.

The pipes of an organ also have harmonics in their vibrating air columns. If the pipe is fat, the strength of the harmonics is low. If the diameter is small compared to the length, the harmonics are strong. Large diameter pipes are called *flutes*, medium diameters *diapasons*, and small diameters *strings*. The pipes also have different shapes (tapered) and some have closed ends. The closed pipes have a node at one end but not the other. This means the fundamental occurs when the length of the pipe is λ /4 and the frequency is one octave (1/2) lower in frequency. The clarinet and flute are approximately the same length but the vibrating air column in the clarinet is closed at one end (reed), which allows it to produce lower notes than the flute. The longest half wave organ pipe in the US is in New Jersey and is 64 ft. long. Assuming the velocity of sound in air is 343 *m/s* can you tell me the frequency of this pipe?

J.   It would have been easy if you had given us the length in meters and not feet. Now we have to convert 64 ft. to *m*.

$$64\,ft\ \frac{12\,in}{ft} \times \frac{2.54\,cm}{in} \times \frac{m}{100cm} = 19.5\,m$$
$$\lambda = 2 \times L = 2\ \times 19.5 = 39\,m$$
$$f = \frac{v}{\lambda} = \frac{343}{39} = 8.8\,Hz$$

I think that is crazy. I remember that the lowest frequency we can hear is about 20 *Hz*. What is the reason for a frequency so low you can't hear it?

Mr. T.   You can't hear it but you can feel it. I have never felt it and I don't know who writes music that contains such low notes. It would be exciting to experience. The lowest note in most church organs is a 16 ft. pipe and is probably only an 8 ft. one with a closed end.

K.   We have switched gears from light waves to string waves and now sound waves. Mr. Tweed could you go over the similarities and differences between these waves.

Mr. T.   Equations 6.1 and 6.2 apply to all these waves but light (electromagnetic waves) and string waves are called *transverse* waves because the thing that is waving has a direction at right angles to the propagation direction. In the case of light the electric part is perpendicular to the magnetic part and both are at right angles to the wave direction. This allows us to have polarized light which has only one vibrational direction. The light waves from the sun are randomly polarized in all directions.

When this light is reflected off a surface like water, the horizontally polarized (electric field) waves are reflected more than the vertically polarized ones. These can be blocked by polarized lenses which allow only vertically polarized light through. The rope waves can also be polarized by changing the direction of the hand vibrations.

Sound waves in air or gas are longitudinal and can't be polarized. The molecules move forward and backward in the direction of propagation. When they move together, a region of high pressure is created. When they move apart, a low pressure region is created.

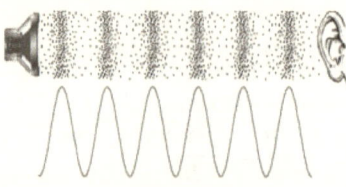

Fig. 6.19

Display of regions of high pressure in a sound wave.

Mr. T.   That's probably enough talk for a while. Why don't you go see Fritz?

In the simulator

F.   I am glad you're back; I was getting bored. Jill and Kay, please pick up your slide whistles and try blowing a note.

J.   We are blowing and it appears that both whistles have been adjusted to the same note.

F.   Yes, I did that. Kay, I want you to make your pitch just a tiny bit higher. When you finish, both of you blow again.

K.   That's interesting, the sound is going whaa whaa whaa.

F.   Those are called beats. The two sounds are going in and out of phase. When they are in phase, the sound is loud; when out of phase, they cancel (interfere).

The frequency of the beats is equal to the difference in frequency of the two sounds.

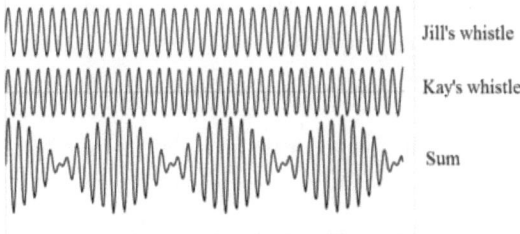

Fig. 6.20
Sound interference or beats between two close frequencies

J. That's cool. Is this phenomenon useful for anything or is it just something to know about?

F. It is not only useful for science but many other things including music. Jill, if you play a 100 *Hz* note (G), what are the harmonics.

J. That easy, the harmonics are 200, 300, 400, etc.

F. Kay, if you play a D which is a fifth higher (150 *Hz*), what are your harmonics?

K. They are 300, 450, 600, 750, and 900 *Hz*. I see what you're getting at; 300, 600, and 900 *Hz* are the same frequency as Jill's harmonics and if my note is out of tune, there will be beats between Jill's and my harmonics in the chord.

F. Yes, it will be a dissonant chord because of these beats. Now I will show you another use of these beats. First let's adjust your whistles to the same frequency.

Jill, there is a small car on tracks. Get in and take the car to the end of the tracks. When you get there, turn around, come back fast, and blow your whistle.

Kay, when she is coming back blow your whistle, don't blow too hard; we want to see if your whistle interferes with Jill's and she will be some distance from us. If your's is too loud, we won't hear the interference.

J.   I am on my way back and sounding my note.

K.   When I blow, I hear the beats. I don't understand; we are sounding the same frequency. How can you have beats when the sounds have the same frequency.

F.   The motion changes the frequency we hear. This is called the Doppler Effect. The following figure shows waves being generated by a moving source.

Fig. 6.21
Picture showing waves from a moving source

Jill, can you tell me which direction the source is moving?

J.   The object is moving down the page; the wavelength in front of it is less than the wavelength behind. The source is trying to catch the wave it just emitted which reduces the wavelength. If the wavelength is changed, the frequency must change according to equation 6.2.

F.   Kay, can you tell me how much the wavelength has been reduced.

K.   I think so. If the velocity of the source is v_s and the new wavelength is λ_n, the source will have moved by $v_s \times T$ where T is the period of the generated wave. The velocity of the wave in the medium is v. Because the period is the reciprocal of the frequency the new wavelength will be given by:
$$\lambda_n = \lambda - v_s / f$$
Using equation 6.2, λ_n and λ are given by:
$$\lambda_n = v / f_n \text{ and } \lambda = v / f$$
After substituting these results into the above relation we get:
$$v / f_n = v / f - v_s / f = (v - v_s) / f$$

This can be solved for the new frequency:

$$f_n = f\left(\frac{v}{v - v_s}\right) \tag{6.3}$$

F. Good job! The Doppler frequency of Jill's whistle is higher than Kay's so we get beats. If we measure the beats we can determine the velocity (v_s) of the car. This is exactly what happens when the police measure your speed with radar.

K. I don't understand; there is no source of waves in the car as there was in our experiment.

F. The police bounce the waves off the car and they come back shifted in frequency due to the speed of the car. The frequencies are very high but the equations are the same ones that you have developed. The Doppler radar combines the transmitted frequency with the frequency shifted return from the moving car in order to create the beats.

   If you work it out, the beat frequency will be close to proportional to the speed of the car. This assumes the speed of the car is much less than the wave speed.

   One last thing, what happens when the car is moving away?

J. The Doppler frequency will be lower. However, I am not sure you can tell by just measuring the beat frequency.

F. Good thinking! The beat frequency is a function of the difference between the Doppler and source frequencies. It doesn't care which is lower. I wonder if anyone has used this as a defense in a traffic court.

K. That would be ridiculous. If you are moving at 70 mph toward or away from the policeman, you are still violating the speed limit.

F. That makes sense. See you next time.

Session 7
Waves, Particles, and Confusion

"It would be possible to describe everything scientifically, but it would make no sense; it would be without meaning, as if you described a Beethoven symphony as a variation of wave pressure." — <u>Albert Einstein</u>

In the classroom

Mr. T.   You are now aware that light is an electromagnetic wave. It turns out that things are not so easy. Today we will examine an effect that was described by **Einstein** and resulted in him being awarded a Nobel Prize. The amazing part of this story was that he published two papers in the same journal issue. The second one was the Special Theory of Relativity, and he could have been awarded the prize for that one as well.

The best way to describe this effect is probably in the simulator. I think Fritz is waiting for you.

In the simulator

F.   I have the apparatus ready to show you Einstein's photoelectric effect. This is not exactly the way he did it; in fact I am not sure he ever did any real experiments. He was a theoretician and others did the experimental work.

The equipment consists of a Cesium metal plate and a metal collector, both in a vacuum bottle with connections to a special voltmeter that requires very little electrical current for a reading. You also have a light source that has adjustable intensity and color.

Place the bottle in its stand, connect the voltmeter, and adjust the light source so that it shines on the Cesium metal plate.

J.   Why is the metal plate made of Cesium? I have never heard of Cesium before.

F. It is a metallic element and has one property that we need for this experiment. It takes less energy to remove an electron from the surface than is required for many metals.

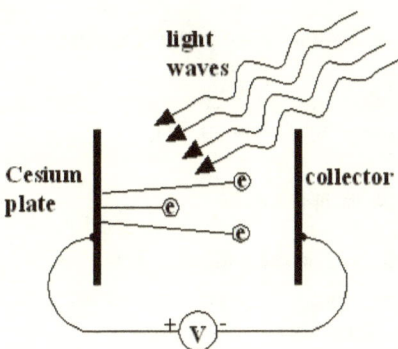

Fig. 7.1
Experimental setup for the photoelectric effect

J. We have it set up. Fritz, what is supposed to happen?

F. If everything works correctly, when the light shines on the plate, electrons will come off the plate and move toward the collector. After a short time, the collector will become charged negatively and we will read a voltage on the meter.

K. What color should I shine on the bottle?

F. You choose anything you like.

K. It is presently set for red, so I will just leave it alone. I turned it on.

J. Nothing is happening! Kay, turn up the intensity.

K. I turned it up and still nothing is happening. Fritz, what is the trouble.

F. You only have one other parameter that can be changed. Try changing the color.

K. OK, I will turn the adjustment to blue light. The instrument tells me that the wavelength is 4.7×10^{-7} *m* and the frequency is 6.38×10^{14} *Hz*. Fritz, it was nice of you to give us both the frequency and wavelength. I could have calculated one or the other from equation 6.2 and the speed of light ($c = 3 \times 10^8$ *m/s*) but now I don't have to.

K. I turned on the light and the voltmeter slowly moved up to 0.54 *V*.

J. I turned up the intensity but the voltmeter did not move. Fritz, why not?

F. It appears that the phenomenon is insensitive to intensity. Turn the light off, discharge the apparatus, and set the light for a new color at very low intensity.

J. How do I discharge the apparatus?

F. Electrons have collected on the collector. If you briefly connect the collector to the Cesium plate, they will travel through the wire and there will be no voltage across the voltmeter.

K. Everything is set. I have selected the color violet ($\lambda = 3.9 \times 10^{-7}m$, $f = 7.69 \times 10^{14}$ *Hz*). The light is on and the voltage is slowly rising. It stopped at 1.08 *V*. I turned up the intensity, but there was no change in the voltage. What should I do now?

F. Try a few more colors, look at the data, and see if you can make any sense out of it.

K. I'll try green now.
    ($\lambda = 5.35 \times 10^{-7}m$, $f = 5.61 \times 10^{14}$ *Hz*) I left the intensity high and after I discharged it, the voltage went up quickly but only to 0.22 *V*.

    Everything was repeated for yellow
    ($\lambda = 5.87 \times 10^{-7}m$, $f = 5.11 \times 10^{14}$ *Hz*). This time the voltage was 0.01 *V*.

    I need to make a chart before I try to make sense out of this.

Color	Wavelength	Frequency	Voltage
Red	7×10^{-7} *m*	4.28×10^{14} *Hz*	0
Yellow	5.87×10^{-7} *m*	5.11×10^{14} *Hz*	0.01 *V*
Green	5.35×10^{-7} *m*	5.61×10^{14} *Hz*	0.22 *V*
Blue	4.7×10^{-7} *m*	6.38×10^{14} *Hz*	0.54 *V*
Violet	3.9×10^{-7} *m*	7.69×10^{14} *Hz*	1.08 *V*

After taking a quick look at the data, I can say that if the light frequency is increased, the voltage increases. I can also say that the voltage does not depend on the intensity of light, and no matter how intense the red light is, the voltmeter does not move.

F. That is not what was expected when they first did the experiment. They expected that as the wave amplitude (intensity) was increased, the electrons would fly off the surface more often with greater energy and the voltage would go up.

J. As more and more electrons are kicked off the surface and collect on the collector, why doesn't the voltage keep rising? More charge will create a greater voltage.

K. I can answer that. If you have a voltage between the plates, there will be an electric field that will push the electrons back toward the Cesium plate. They will need more kinetic energy to make the trip all the way to the collector.

J. I get it; changing the color changes the kinetic energy the electron has when it is ejected from the surface.

K. The light wave evenly illuminates the whole surface but the electrons seem to come from random spots. Does the wave concentrate all its energy on one spot for an instant to send the electron flying and randomly choose positions all over the plate as long as the light shines?

F. I am sure you have many questions as did the scientists at the time these experiments were performed. Let's plot your data before we try to understand Einstein's speculation about what was going on.

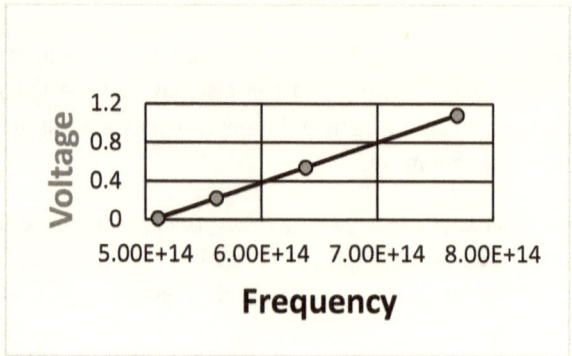

Fig. 7.1

Graph showing the voltage in the photoelectric tube produced by light of different frequencies.

K. I see that four of our points lie in a straight line which means that the mathematic relationship between them will be simple. We just need to find the slope and intercept.

F. I will do the math for you; the equation of the line of your data is.

$$V = 4.1 \times 10^{-15} f - 2.1$$

J. Why does the line start at -2.1 *V* and not at zero?

F. The energy required to remove an electron from the Cesium surface is 2.1 *eV* which means nothing happens until the light has enough energy to remove an electron from the surface. If no energy was required to kick an electron off the surface, the voltage would rise according to the following relation:
$$V = 4.1 \times 10^{-15} f$$

K. A Volt is a Joule per coulomb. If we want to know the total energy given to the electron by the light in fundamental units we should multiply the voltage by the charge on an electron (*e*).
$$(e = 1.6 \times 10^{-19} \ C)$$
$$E = (4.1 \times 10^{-15} \times 1.6 \times 10^{-19}) f = 6.6 \times 10^{-34} f$$

F. If you want to know something amazing, the constant 6.6×10^{-34} *J / Hz* is the same constant found by **Max Plank** in his work on radiation from hot objects.

It is called Plank's constant and is designated by h. The relation that Einstein has given us is:

$$E = hf \qquad (7.1)$$

J.   I still can't make a lot of sense out of this experiment and I am sure that it took Einstein some time to figure it out even after he had the data. Surely you don't expect us to understand its significance. I need some help.

F.   I expect you to understand but only after I help with Einstein's explanation. You are absolutely correct; many people were aware of the photoelectric effect but it took Einstein to give us an explanation. I will try to give you some sense of his ideas.

According to Einstein, the light shining on the plate is composed of quantized bundles of energy. These quanta strike the plate and liberate electrons. The energy of each of these quanta is given by $E = hf$. The phenomenon we are seeing can't be described by simple waves. The energy of a wave depends on its amplitude and the amplitude is not quantized. The name we use for these quanta was given to them years after Einstein's paper; we call them photons and tend to think of them as particles of light.

Thinking of them as particles is easier if their frequency is very high (Gamma rays). Gamma rays hit things, destructively ionize molecules, and generally behave like particles. Low frequency radiation like radio waves don't have easily observable particle properties and we think of them mainly as waves.

J.   Just a minute, you said Gamma rays have a very high frequency and I assume a short wavelength. How is it that we describe the properties of these rays using wave properties and think of them as particles? Are they waves or particles?

F.   This is one of the fundamental mysteries of physics. There is no answer to this question and as we shall see later, the situation becomes weirder and weirder. In some situations we will describe light as waves and in other as particles.

In the classroom

Mr. T.   You should now be confused. Light waves act like particles as well as waves. That is strange. The photoelectric effect shows particle behavior and the many interference effects indicate wave behavior.

J.   What are these interference effects? The only one we have examined is the formation of beats. We looked at standing waves of sound but not electromagnetic standing waves.

Mr. T.   We will have a chance to examine a number of different interference effects but the double slit experiment may shed the most light on the phenomenon of interest now. (Pun intended)

Historically, it is difficult to find a simple physics experiment that is referenced or written about more times than the double slit one. It offers a window into one of the most philosophically difficult problems in physics. Therefore, I will start the explanation of the experiment in the classroom before sending you into the simulator for an experiment.

Figure 7.2 shows a picture of what happens when a periodic plane wave passes through two small openings at the bottom of the figure. After leaving the openings or slits, the waves spread out and interfere with each other. Destructive interference occurs when a crest from one slit arrives simultaneously with a trough from the other slit. At a position where the distance to each slit is identical the waves will reinforce. Any position where the distance to one slit is different by a multiple of a wavelength will also show reinforcement.

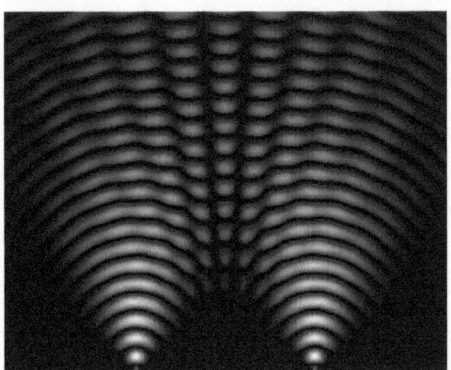

Fig. 7.2
Snap shot of the waves from a double slit interference pattern simulated by a computer.

As the waves proceed, the bright spots in the middle, which represent reinforcement, move up. The vertical dark bands represent nodes (interference) and do not move. Any point on a node will be an odd multiple of a half wavelength ($\delta = \lambda / 2, 3\lambda / 2, 5\lambda / 25, 7\lambda / 2$, etc.) farther from one slit than the other. If the point on the node is far from the slits, we can use the following diagram to help with the math. The symbol δ is used to designate the extra distance one wave must travel.

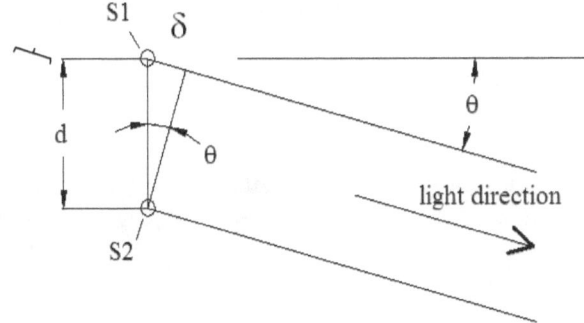

Fig. 7.3
Geometry for a double slit experiment. The slits are represented by S1 and S2.

The two rays from the slits are shown as parallel lines because the node position is far away. Kay, what can you get mathematically from the diagram?

K.   I see a triangle with two of its sides and the angle labeled. I also see that if I pick up the triangle and rotate it by 90 degrees, it will have the same angle as the other one in the diagram. The trigonometric relation for this triangle is:

$$\sin \theta = \delta / d \qquad\qquad (7.2)$$

Mr. T.   We now need to expand the diagram in order to find another triangle with the angle θ.

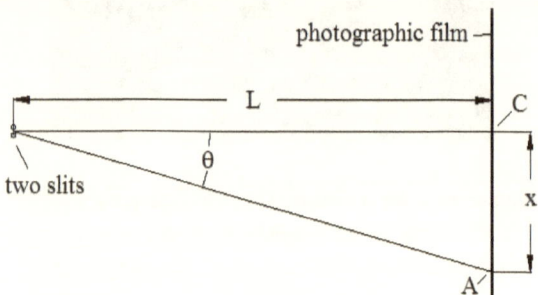

Fig. 7.4
Geometry of a double slit experiment at a greater distance from the slits.

Jill, can you make sense out of this diagram?

J.   I see clearly the second triangle with the angle θ. However, I can't write a relationship for the sine of the angle. I can write:

$$\tan \theta = x / L \qquad\qquad (7.3)$$

However, I can see that L is the distance from the slits to the photographic film, but what is x?

Mr. T. x is the distance from the center of the pattern to the point A on the pattern. If we limit our experiment to very small angles, we can use the approximation:

$$\tan \theta \sim \sin \theta \quad \text{for small angles} \qquad\qquad (7.4)$$

J.   I don't know how you can do that.

Mr. T.   I will give you an example. Let's use one degree.

$$\tan (1) = 0.017455 \sim \sin (1) = 0.017452$$

You can see that they are very close and I suggest that you keep track of the angles in the example that I describe and check our approximation.

We now have two expressions for the sine or tangent of the angle; what should we do now?

K.   I will set them equal to each other.

$$x / L = \delta / d \qquad (7.5)$$

If the point A is on the first node, the difference in distance (δ) must equal $\lambda / 2$. If this is substituted into equation 7.5, an expression for λ can be obtained.

$$\lambda = \frac{2xd}{L} \qquad (7.6)$$

This means we can measure the wavelength of light if we can determine the slit separation (d).

Mr. T.   Let's try; I have a picture of an optical double slit diffraction pattern.

4.2 cm

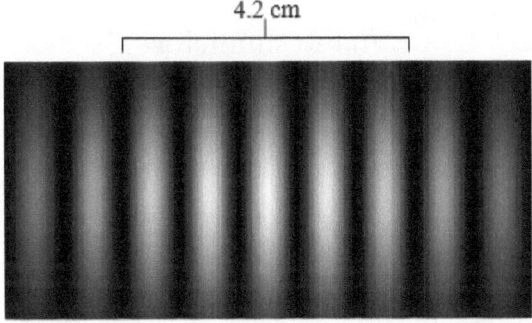

Fig. 7.3

Photo showing the pattern made by the light from a double slit experiment striking a film.

The slits were made by holding two thin razor blades together and scratching the black smoked surface of a microscope slide. d was determined by measuring the thickness of ten razor blades and dividing by ten. $d = 0.16$ *mm* or 1.6×10^{-4} *m*. L was 1.9 *m*. x was determined by finding the distance between nodes and dividing by 2. $x = (4.2/5)/2 = 0.42$ *cm.* Jill, why don't you do the calculation for λ?

J.   I first need to make sure I have everything in meters. The only number I have to change is 0.42 *cm* which is 0.42×10^{-2} *m*. After substituting into equation 7.6, I get:

$$\lambda = \frac{2 \times 0.42 \times 10^{-2} \times 1.6 \times 10^{-4}}{1.9} = 7.1 \times 10^{-7} \ m$$

Mr. T.   You can't see the color in the picture but now you can tell me the color.

J.   The color is red; I learned that from the photoelectric experiment. I also did something else you asked us to do. I kept track of the angle. The sine of the angle would be x / L which allowed me to calculate the angle θ as 0.12 degrees. θ is so small that setting the tangent equal to the sine seems ok.

K.   Mr. Tweed, I understand what we are doing but you indicated that this experiment was really important; it just seems like another interesting phenomenon. Why is it so important?

Mr. T.   Go and see Fritz; he will perform this experiment again in a slightly different way and you will begin to understand its importance.

In the simulator

F.   I am glad to see you back. I have the slit experiment all ready and the set-up is shown below.

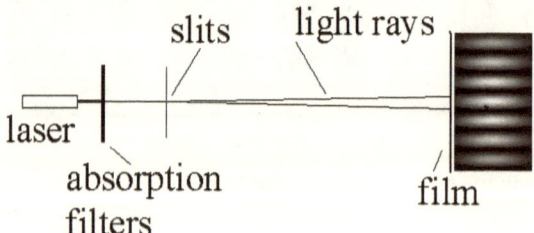

Fig. 7.4

General features of the double slit experiment

The distance from the slits to the film is the same (1.9 *m*) and all the other features are the same as discussed with Mr. Tweed. However, absorption filters have been added to weaken the light beam.

J.   Why do you want to make the light weak? We will not be able to see it.

F.   I don't expect you to see it but we will have a photographic record of what happens. Do you remember what you learned about Einstein's photoelectric effect?

J.   Einstein showed that light waves acted like particles which we now call photons and he gave us a formula for their energy. $E = hf$

F.   Good! Now we want to examine the double slit experiment and see if we can see what happens with these photons or light particles. In order to do that, we need to reduce the number of photons down to the point where there is only one at a time in the experiment.

K.   How can we be sure that there is only one at a time?

F.   First, let's calculate how many photons are in the beam. The laser puts out a maximum of 5 *mW* which is $5x10^{-3}$ *J / s*. Kay, how many photons per second is that?

K.   Each photon has an energy of *hf* and we learned that red light had a frequency of about $4x10^{14}$ *Hz*. Plank's constant is $6.6x10^{-34}$ *J / Hz*. Therefore, the energy of each photon is about $2.6x10^{-19}$ *J*. If I divide the power of the beam by this number I will get the number of photons per second.
$$N = 5x10^{-3} / 2.6x10^{-19} = 2x10^{16} \text{ photons per second}$$

J.   That is twenty million billion. How can we ever reduce that to only a few per second?

F.   We don't have to reduce it that far; we only have to be sure that only one at a time are in the apparatus.

K.   I don't understand why we want only one at a time?

F.   If we have only one at a time, we expect it to go through only one slit. If it only goes through one slit, should we see an interference pattern?

J.   The pattern comes from the interference of two waves. If the photon only goes through one slit at a time, I don't think we will see an interference pattern.

F. Let's try to reduce the number of photons down to about 20 per second. It takes very little time to traverse the apparatus so I would definitely not expect two at the same time. First, the slits have very little area, so not more than 1% of the beam will get through. It is also possible to buy absorption filters that will reduce the beam by a factor of 10^2. Kay, how many filters do I need?

K. We want to reduce 2×10^{16} photons per second to 20 per second. That is a factor of 10^{15}. The slits cut the number by 10^2 (equivalent to one filter) so if we attenuate using 7 more filters, we will be down by a total factor of 10^{16} which gives us 2 photons per second. If we use 6 filters we will have 200 photons per second. I think 6 should be enough.

J. If we are going to attenuate the beam that much, we have to take extreme measures to avoid having some light sneak around the filters and spoil the experiment.

F. That's correct but don't worry, this is a simulator. In the real world it would also be difficult to find a film that was sensitive enough for our simple experiment but we don't have that problem.

    Are you ready to start? How long should we expose the film?

J. I will turn the light on for 0.1 seconds. That will give us just a few pixels

    Fritz, will you develop the film?

F. I don't have to, we use digital photography and the word film only refers to the photosensitive screen.

Fig. 7.5
Pattern after a short exposure

J. I will try a little longer.

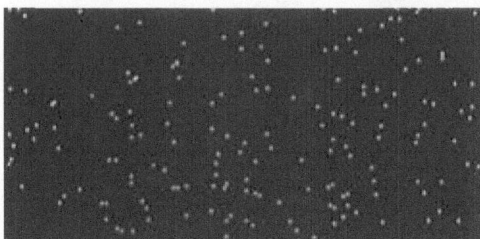

Fig. 7.6
Pattern after about a second of exposure

K. It looks random. I don't think we will see any interference pattern with only one photon at a time. Jill, give it a blast this time.

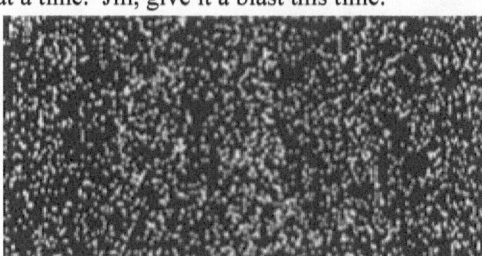

Fig. 7.7
Pattern after a few seconds

J. I still don't see anything, should I give it more? Why not?

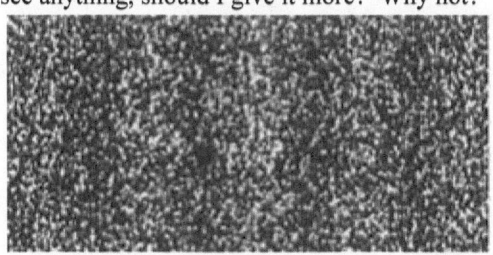

Fig. 7.8
Pattern after several thousand photons

J. I think I see something. I'll give it one more blast.

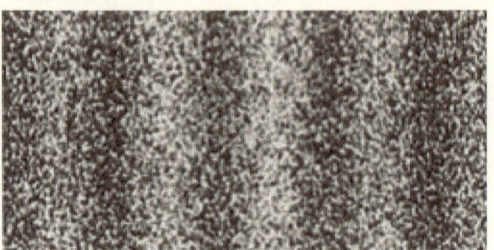

Fig. 7.9
Final pattern after many thousand photons

K. Wow! We do get an interference pattern even with only one photon at a time. Fritz, what does it mean?

F. Unfortunately, I can't give you a simple answer; I can't even give you a complicated answer that makes a lot of sense. It is possible to calculate wave amplitudes that tell us the probability of finding a photon at particular positions. We only get probabilities, never any exact answers.

A photon seems to be just a wiggle in the electromagnetic field of charged particles. However, that is not the whole story. Photons also have momentum. The fact that light had momentum was known long before the idea of photons. It was found both theoretically and by experiment that a beam of light had a momentum equal to its energy divided by the speed of light.

$$p = E / c \qquad (7.7)$$

Kay, can you use Einstein's energy formula to write the momentum relation for a single photon?

K. That's easy, I will just substitute *hf* for *E*.

$$p_{photon} = hf / c \qquad (7.8)$$

I might even be able to simplify it a little more because $\lambda f = c$.

$$p_{photon} = h / \lambda \qquad (7.9)$$

F. That's what I was looking for. Light beams or photons don't have very much momentum but they do have some. Along with the solar wind they are responsible for pushing the tail of a comet away from the sun. When the wavelength gets very short as it does with gamma rays, we begin to see distinct effects of the photon's momentum.

J. What kind of effects can we see?

F.   Well, we won't see them with the naked eye, but when a gamma ray (photon) strikes an electron we can see (using the right instruments) the electron move as if it had been hit by a solid particle. Gamma rays can also move through our bodies ionizing molecules and sending electrons flying. Exposure to very much high frequency radiation is not good.

It's time to end this simulator session but I think I will see you again very shortly.

In the classroom

Mr. T.   I hope you enjoyed learning about the photon's momentum. Before we move on I want to point out a couple of details so you won't be confused if you read some other material on this subject. Most scientists use the symbol c for the speed of light and v (nu) for the frequency (f) of light. We, however, will continue to use f.

Our next topic involves the work of a French physicist by the name of De Broglie. Sometime after Einstein published his work on relativity and the photoelectric effect, De Broglie looked at the relation for the photon's momentum rewritten to show a solution for the wavelength.
$$\lambda = h \, / \, p \tag{7.10}$$
His logic was much more involved but he essentially said maybe this relation works for everything. This would mean that anything that had momentum would act like a wave and have a wavelength that depended only on its momentum.

J.   Does that mean if I were moving, I would have a wavelength?

Mr. T.   That's exactly what it means. However, go ahead and figure out your wavelength.

J.   My mass is about 60 kg and if I move at 1 $m \, / \, s$, my momentum would be 60 $\frac{kg \, m}{s}$. Plank's constant is 6.6×10^{-34} Js, so my wavelength would be:
$\lambda = 6.6 \times 10^{-34} \, / \, 60 = 1.1 \times 10^{-35}$ m. That is so small I have no way to even think about it.

Mr. T.   Now you understand why the idea of a De Broglie wavelength is only useful for very small things.

K.   Mr. Tweed, when you say small, how small do you really mean?

Mr. T.   If we consider electrons, the De Broglie wavelength will be somewhat like the size of an atom. If we consider protons or neutrons, the wavelength will be comparable to nuclear sizes. Photons as we have seen can have wavelengths of almost any size.

K.   Are you saying that De Broglie was correct? What evidence do we have that particles act like waves?

Mr. T.   Actually, we have a huge amount of evidence. However, it is clear that it will be difficult to see evidence of wave properties for large objects. Jill is good example. She is correct, her wavelength is much too small to even think about. We have the highest probability of success with particles with small mass that are not moving too fast. Their momentum must be low in order for their wavelength to be large.

Let's look at some of the evidence that electrons have wave properties.

J.   What kind of wave properties are we talking about?

Mr. T.   If we have waves, we should see interference effects. If we see interference, we have evidence for waves. I think Fritz can show you.

In the simulator

F.   I told you that you would be back soon. We will now examine one technique that demonstrates that electrons behave like waves.

K.   Before we do that, I would like to know generally what the wavelength of an electron is. I don't know what its mass is and I don't know how fast it moves or its momentum. Can we measure these parameters?

F.   Yes we can make these measurements. Before we start let me explain an electron gun. These are used in all cathode ray tubes for computers and TVs. The cathode is usually a small metal tube that is covered with a coating that make it easy for electrons to escape the surface. Inside the tube is a heater which raises the temperature until the cathode is glowing brightly. The electrons will now be moving about very rapidly and many will escape the surface and fly into the vacuum surrounding the cathode. An anode with a positive voltage attracts these electrons and they are accelerated by the electric field until they hit the anode. There is a hole in the anode and some of the electrons pass through forming a beam (cathode ray).

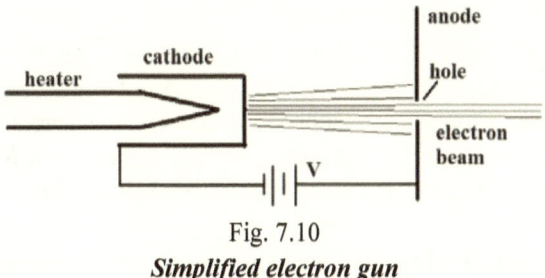

Fig. 7.10
Simplified electron gun

Many design details have been left out. The efficiency is optimized by forcing a high percentage of the electrons to go through the hole and features are added to focus the beam after it leaves the hole.

The electrons in the gun I have provided for you will produce a beam of electrons that have been accelerated using 66 *V* of electrical potential. You should be able to calculate their kinetic energy.

J. I can do that. A Volt is energy per Coulomb of charge, so if I multiply 66 by the charge on an electron, I will get the kinetic energy of the electron. *KE* = $66 \frac{J}{C} \times 1.6 \times 10^{-19}\ C = 1.056 \times 10^{-17}\ J$

F. We know the *KE* of the electrons and now we can measure their momentum. Remember from session 5 we found that all we had to do to determine momentum was to measure the radius of the flight path in a known magnetic field.

The information we need is in equation 5.3:

$$p = BqR$$

In this case *q* is the charge on an electron *e*. I have produced a vertical magnetic field (*B*) in the room equal to $1 \times 10^{-4}\ T$. The *T* stands for Tesla.

J.   We can't see electrons. How are we going to measure the radius of their path?

F.   At present we are in a vacuum. I will add just a tiny bit of gas so that the gas molecules will give off light when the electron beam ionizes them; the path will be illuminated. Kay, please take a ruler and measure the radius. I will turn off the lights now so that we can see the beam.

K.   I see the beam but just barely see the ruler. The beam is moving in a circle. I can't measure the radius because the center has no mark, but I can get the diameter. $D = 0.55\ m$
I can calculate the momentum now.
$$p = 1\text{x}10^{-4} \times 1.6\text{x}10^{-19} \times 0.55/2 = 4.4\times10^{-24}\ kg\ m/s$$

F.   You now have enough information to calculate the wavelength, but you asked about the mass. Let's figure that out. We have values for the KE and p. You should be able to calculate the mass from these values. Who wants to try?

K.   I think I can do it. $KE = mv^2/2$ and $p = mv$. I have two equations and two unknowns. I need to remove v. I will solve the momentum equation for v, substitute it into the energy equation, and solve for m. $v = p/m$ which after substitution into the energy equation gives: $KE = p^2 / 2m$
Solving for m gives: $m = p^2 / 2KE$ Therefore, the mass of an electron is:
$m = (4.4\text{x}10^{-24})^2 / (2\text{x}1.056\text{x}10^{-17}) = 9.1\text{x}10^{-31} kg$

J.   I think I can calculate the wavelength. $\lambda = h/p$ or
$$\lambda = (6.6\text{x}10^{-34}) / (4.4\text{x}10^{-24}) = 1.5\text{x}10^{-10} m$$

K.   If we compare that number to the wavelengths we were using for visible light the electron's wavelength is about three thousand times smaller. If we are going to use slits to get an inference pattern they need to be very close together.

F.   We will not use slits, we will use the regular arrangement of atoms on the surface of a Silicon crystal. The atoms are a little more than $2\text{x}10^{-10} m$ apart. You can see that the separation is only a little larger than a wavelength. A diagram of the apparatus is shown below.

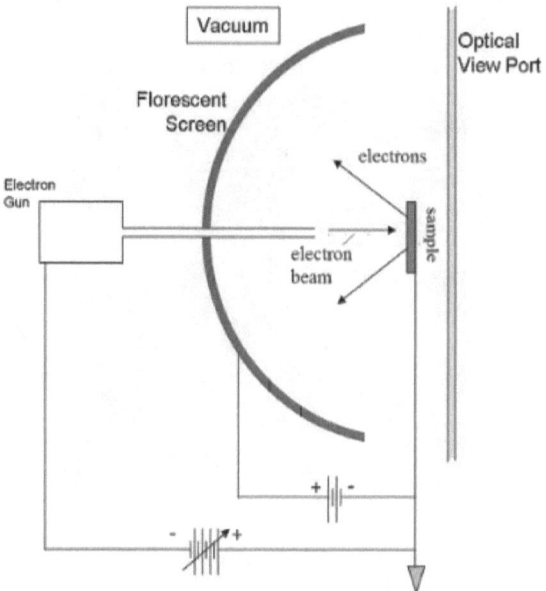

Fig. 7.11
Low energy electron diffraction apparatus

The electrons are fired at the Silicon sample which is at the center of a fluorescent screen. A typical picture as seen through the view port is shown below.

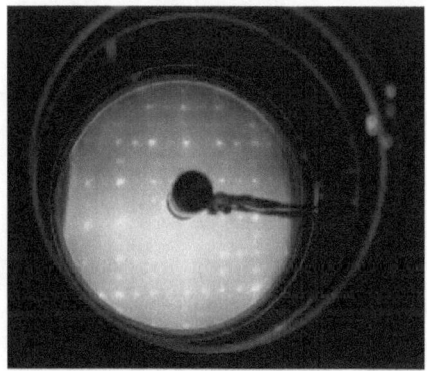

Fig. 7.12
LEED picture –Wikipedia

F.   We have our own apparatus. Place your gun in position and see what you get. I removed the gas from the room; you will no longer see the beam.

J.   I put the gun in place and it is still set for 66 *V*. We get a very nice pattern.

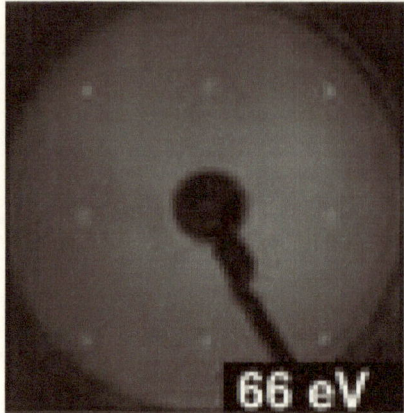

Fig. 7.13
66 V LEED picture of Silicon

F.   Try increasing the voltage and see what happens to the diffraction pattern.

Fig. 7.14
244 V Silicon diffraction picture

J.   We increased the voltage by almost four times and the pattern only decreased by about a factor of two. Why so little?

K. The momentum is not a direct function of the energy (voltage). You can see that from their defining equations: *p=mv* and **KE=mv²/2** If we put these together and solve for *p*, we get:

$$p = \sqrt{2m\,KE}$$

This shows that if we increase the energy by 4 the momentum will increase by 2 and the wavelength will decrease by 2.

Let's increase the voltage some more.

Fig. 7.15
Silicon LEED pattern at 450 V

F. Who can tell me why the fluorescent dot pattern is square?

J. It must be because the atoms of Silicon are arranged in a square pattern. The crystal must be oriented so that the face is perpendicular to the beam. The crystal is probably cubic.

F. Let me show you one that is not cubic.

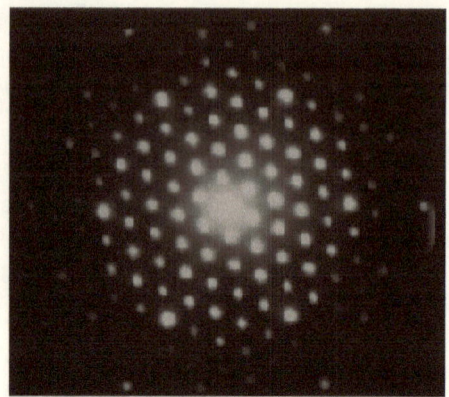

Fig. 7.16
LEED picture of an unknown crystal

What structure would you guess for this sample?

K. It looks hexagonal. It appears that we are only seeing the surface layer of atoms. Why do we not see the electron wave reflections from the interior of the sample?

F. This technique is called Low Energy Electron Diffraction (LEED) and low energy electrons will not penetrate through the surface of most materials. There are difficulties in looking at surfaces because most pure materials are very active chemically on the surface. For instance, you will never see the pure surface of Aluminum; you always see Aluminum Oxide covering the surface. The surfaces in LEED studies must be prepared in a vacuum. If there is any contamination, the pictures will not be good.

J. Are there other applications that use the wave nature of electrons?

F. Yes there a number of diffraction experiments that we don't have time to describe here, and there are also several types of electron microscopes.

I think we are done today. See you next time.

Session 8

Atoms

It is often stated that of all the theories proposed in this century, the silliest is quantum theory. In fact, some say that the only thing that quantum theory has going for it is that it is unquestionably correct. --Michio Kaku

In the classroom

Mr. T.   Today we will attempt to say a few things about atomic structure. In the early part of the twentieth century, many thought that a positive nucleus was surrounded by orbiting electrons in a similar way to the planets orbiting the sun. There are several problems with this model. The first involves radiation of the orbiting electron. Kay, can you tell me about radiation from an electron moving in a circle?

K.   I know that when you accelerate an electric charge it will radiate and any object moving in a circle is accelerating. I don't understand why this presents a problem with the orbital model of the atom.

Mr. T.   If the electron radiates it must lose energy. If it loses energy, what will it do?

K.   It would slow down and spiral closer to the nucleus. I think I see the problem; this would cause the atom to collapse. What keeps this from happening?

J.   We just learned that the electron acts like a wave. The wavelength is inversely proportional to its momentum ($\lambda = \frac{h}{p}$). If the electron loses energy, its momentum must decrease, and wavelength increase. If a wave is confined to a small space, it will exist as a standing wave and the space can never be smaller than half a wavelength. How can the electron's wavelength get bigger and move closer to the nucleus and into a smaller space? If it moves into a small space, it must have a small wavelength. If it has a small wavelength, it must have a large energy.

Mr. T.   It, therefore, becomes impossible for it to lose energy; it can't spiral into the nucleus. <u>If the electron did not have wave properties, we would not have stable atoms and our world could not exist.</u>

We have been able to examine some wave properties by looking at interference phenomena. In our sessions, we have seen standing waves on strings and in air with pipes, and have been able to do some calculations with regard to the position of nodes and antinodes. Unfortunately, we have not been able to calculate the amplitudes of these waves or their general features other than their nodes and antinodes. This would require the use of wave functions and mathematical techniques far beyond our capabilities. We are capable, however, of understanding some of the fascinating ideas involved.

Early in the development of quantum mechanics a physicist by the name *Schroedinger* developed a wave equation which can be used to calculate the standing waves of electrons around the positive nucleus of an atom. These waves are described by *wave functions* which are awesome to examine.

Before we look at any of these atomic waveforms, let's examine some standing waves that are easier to comprehend. We saw that one dimensional standing waves could exist as harmonics on strings. The string could be divided into *n* parts each of which was equal to half a wavelength. The frequencies were all multiples of the fundamental or lowest harmonic. A similar situation existed with air in pipes.

Two dimensional standing waves are more complicated. Fortunately, a man by the name of **Chladni** developed a method for observing the standing waves in two dimensional plates. He sprinkled fine sand on the plates and excited them in various ways to vibrate. A typical way was to use a violin bow and draw the bow carefully across different portions of the edge of the plate. Waves would move in various directions and form standing waves at different resonant frequencies of the plate. The sand would move to the nodes of these resonances and make visible patterns. He could also create sound at specific frequencies near the plate. When the sound was the same as the natural resonance frequency of the plate, the plate vibrated and the pattern would form. Some examples for a square plate are shown below. Note that the plate is mounted in the center so this becomes a node automatically.

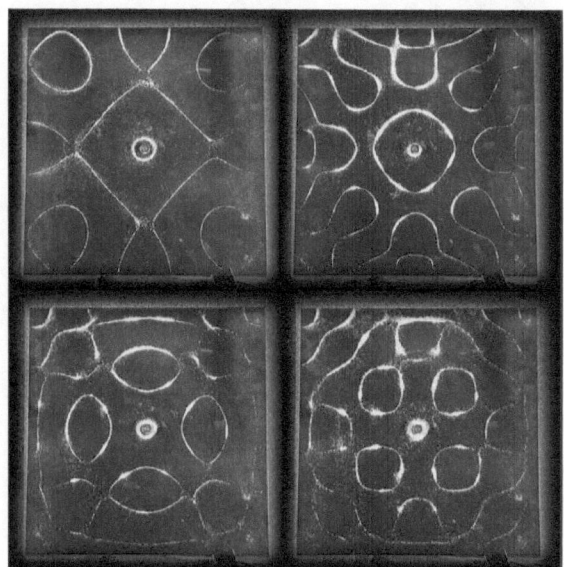

Fig. 8.1
Chladni figures on a square plate

Jill, can you tell me which plate is resonating at the lowest frequency?

J.   That's easy. The lowest frequency will have the longest wavelength and the distance between nodes will be the greatest. The square at the upper left fits this criterion.

Mr. T.   The patterns shown in figure 8.1 are symmetrical in both the x and y directions. I am going to show you some patterns that have different kinds of symmetry.

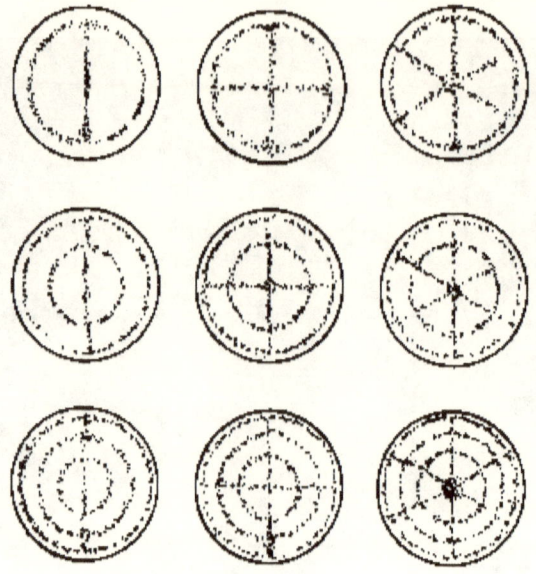

Fig. 8.2
Chladni figures on a circular plate

Kay, look at the upper left pattern and tell me what you see going on there.

K. I see a node running vertically through the pattern and also a circular node around the edge of the plate. I assume that when the part of the plate to the left of the vertical node is moving up, the part to the right would be moving down and vice-versa. I don't understand what is happening around the circular edge; it looks like the plate is clamped all around the outer edge.

F. I think you are correct about the clamping because none of the nodes proceed to the edge in the same fashion that they do with the square plates.

Jill, would you like to explain what is going on with the middle plate?

J. First, I think you are wrong about the clamping. If you look at the pattern on the upper right, you will see some of the radial nodes extending all the way to the edge. The plates are clamped all around the outer edges. I don't know why we can't see this extension in the other patterns, but I'll bet the nodes are there. The radius of the circular node in the upper left pattern is also less than any of the others. It is clearly a node and not just clamping.

The middle pattern is too difficult to just talk about. I am going to label the antinodes as either up or down. I understand that one half period later in time the labeling should switch from up to down. It's interesting that each time you cross a nodal line the motion direction changes. If both sections of the plate moved in the same direction, there couldn't be a node; both sides would move up and the sand would move. I have arbitrarily extended the radial nodal lines and labeled the motion at the edges.

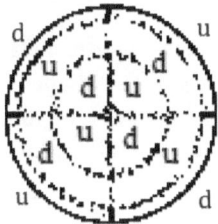

Fig. 8.3
Jill's labeling of the antinodes

In this case as you increase the angle around the circle, you encounter four nodes. It is interesting that you would not encounter an odd number of nodes around the circle as this would mean adjacent antinodes would be moving in the same direction. Mr. Tweed, can these patterns be described with wave functions?

Mr. T.   Yes they can and don't worry, we are not going to derive them. However, we can say a few words about these functions. When we looked at standing waves on a string, we saw that the string could be divided into n parts where n is an integer. Kay, can we use only one integer to describe the resonance patterns on these circular plates?

K.   Maybe we could just count the anti-nodes. I will try. Going from left to right and then down they are:

4	8	12
6	12	18
8	16	24

I can see a problem with this method of classifying the pattern. Some of the numbers repeat and don't differentiate between different patterns; the number 8 could indicate either the lower left or the upper middle of our grouping. I think we need two integers. One could indicate how many anti-nodes are encountered as the angle increases going around the circle (n) and the other (*b*) to indicate the number of anti-nodes encountered as the distance from the center is increased. The pair *n,b* would indicate the pattern and the designations for the nine pictures would be:

2,2	4,2	6,2
2,3	4,3	6,3
2,4	4,4	6,4

I assume *n* and *b* would appear in the mathematical formulations of the resonant patterns.

Mr. T.   Yes and I think we are now ready to consider the resonant wave patterns for the simplest of all atoms, the Hydrogen atom. For a one dimensional string we needed one integer and for a two dimensional plate, we needed two integers. Jill, how many integers do you think we will need for the three dimensional atom?

J.   Mr. Tweed, you have made it too easy. If nature follows the progression one integer for each dimension, then we need three. However, I don't really know why.

Mr. T.   You are correct, it is three, and knowing why would be good but it is perhaps more important that the number three feels right. This is especially true because we don't have enough mathematical tools available for a good explanation.

The letters *n*, *l*, and *m* are used to designate the integers, known a quantum numbers, used for electron atomic wave functions. Later we will look at what each letter stands for but right now let's just look at simplest electron probability densities for changing *n*.

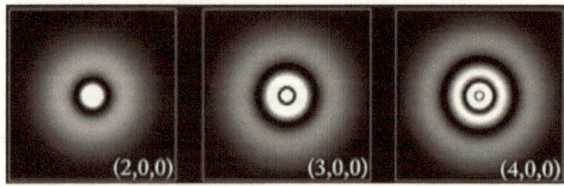

Fig. 8.4

Cross-sections of Hydrogen electron probability densities (orbitals) where the quantum numbers l and m are both zero

Note that these pictures are not to scale; as *n* increases, the diameter becomes larger as you can see in figure 8.5. You can also see that the electron density is zero at the center (nucleus).

Hydrogen Orbital Electron Density

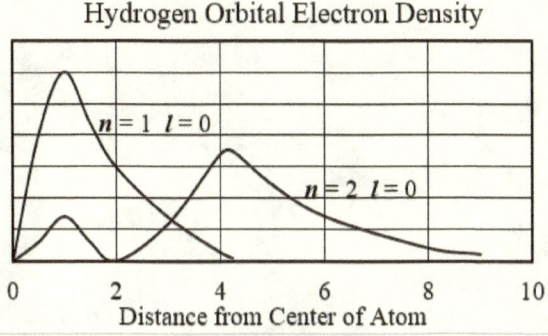

Fig. 8.5

Graph showing how the electron density's radial distance increases with n.

Mr. T.   All the functions shown in figure 8.4 have no angular dependence and look the same in all directions. If they were viewed from outside and not as cross-sections, they would just look like balls.

Kay, tell us what you see in figure 8.4

K.   I assume that the bright areas indicate high probability density (anti-nodes) and the dark areas are nodes or at least where nothing is happening. As you move from the center outward, the number of anti-nodes is equal to *n*.

You indicated that all these pictures were cross-sections and looked the same from any point of view. Therefore, the bright anti-nodes must be hollow balls. What do the orbitals look like when *l* is larger than zero?

Mr. T.   Before I show you some of the other orbitals (probability density patterns), there are some rules about the values of *n*, *l*, and *m* that you should be aware of.

n can be any positive integer.
l can be 0 or any positive integer less than *n*.
m can be 0 or any positive or negative integer less than or equal to *l*.

Example: If *n* is 2 and *l* is 0, *m* must be 0.

If *n* is 2 and *l* is 1, *m* can be -1, 0, or +1.
If *n* is 3 and *l* is 2, *m* can be -2, -1, 0, +1, or +2.

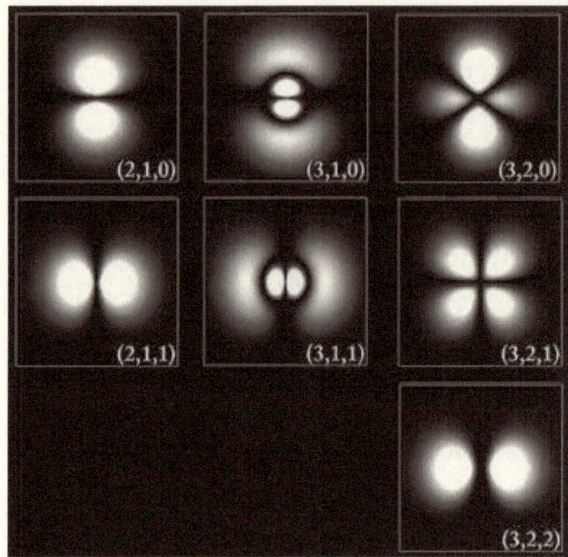

Fig. 8.6
Hydrogen orbital cross-sections for n = 2 and 3

It may be helpful to see three dimensional renditions of these same orbitals. Think of rotating the figures in figure 8.6 about a vertical axis to obtain the 3d renditions in figure 8.7.

$$n,l$$

2,1   3,1   3,2

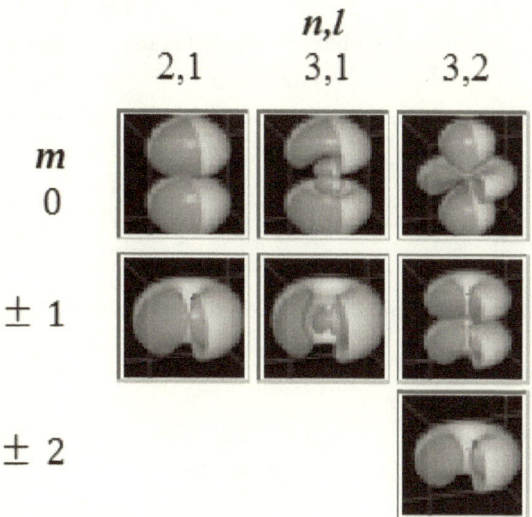

m

0

± 1

± 2

Fig. 8.7

Three dimensional Hydrogen orbitals. The orbitals have been cut open to provide greater clarity.

K. All these calculations of waves give us very interesting pictures but what evidence do we have that any of this is correct?

Mr. T.   We have a lot of evidence but the most important comes from the fact that each orbital has an energy which can be calculated from the Schroedinger wave functions. When an atom of Hydrogen is ionized, the electron is torn lose from the nucleus. If it or another electron is attracted to the nucleus, it doesn't just spiral down. It jumps from one orbital to another giving up energy with each jump. The energy it gives up in the form of a photon is equal to the difference between energy levels involved in the jump. The energy levels are mainly dependent on the quantum number n. A very simple relationship between the orbital energy and n has been calculated:

$$E \propto \frac{-1}{n^2} \qquad (8.1)$$

K. Why is the energy negative?

Mr. T.   The electron is presumed to have zero energy when at a great distance from the nucleus ($n = \infty$) and it requires energy to take it away from the nucleus. Therefore, it must have negative energy for small n.

Einstein has given us the relationship between the energy of a photon and its frequency in equation 7.1 which we studied in the last session.

$$\Delta E = hf \tag{8.2}$$

The delta symbol was used to emphasize that the photon energy depends on the energy change (loss) in a jump from one orbital to another.

Transition of n	3→2	4→2	5→2	6→2	7→2
Name	H-α	H-β	H-γ	H-δ	H-ε
Wavelength (nm)	656.3	486.1	434.1	410.2	397.0
Color	Red	Cyan	Blue	Violet	(Ultraviolet)

Table 8.1
Hydrogen spectra

Hydrogen Emission Spectrum

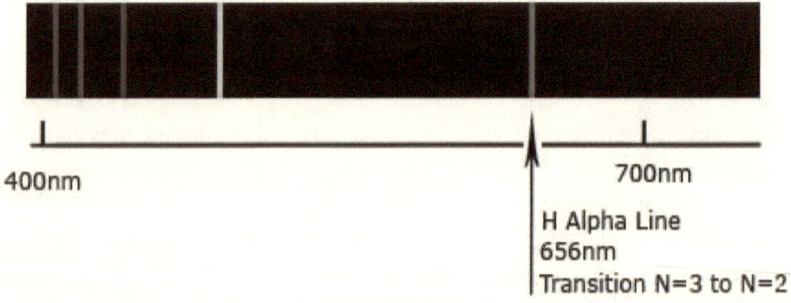

Fig. 8.8
Hydrogen atomic spectra (Balmer series)
The light coming from ionized Hydrogen was sent through a slit in an optical spectrometer that bends short wavelengths more than longer ones

Unfortunately, our eyes can only see light in about one octave of the electromagnetic spectrum. All the transitions (jumps) to the $n = 1$ orbital are in the ultraviolet and invisible. The transitions to the orbitals with $n > 2$ are in the infrared and also invisible. However, Schroedinger wave mechanics theory predicts all these wavelengths. Even the wavelengths invisible to our eyes can be measured using the appropriate instruments.

J.   Are you saying that the electron will not remain in any of the orbitals you have shown us?

Mr. T.   That's correct. The electron will make its way to the 1,0,0 orbital which is called the ground state because it has the lowest energy level. It's like a ball bouncing down the stairs; it could go one step at a time or skip steps.

J.   Mr. Tweed, I think I understand the energy levels but I grew up thinking the electron moved about the nucleus in an orbit like the earth moves about the sun. Even today, I see pictures of electrons orbiting the nucleus and you even called these pictures of the electron density orbitals. It seems that this idea I was taught is not completely dead. As I understand from what you have said, the electron probability densities only tell us the probability of finding an electron; they don't tell us exactly where it is or what it is doing. Let's try to keep this simple. Can you just tell me what the electron is doing in its lowest energy state?

Mr. T.   Unfortunately, I can only say a few things. We know that in the ground state the electron has no orbital angular momentum. That means it is not revolving around the nucleus. However it must be moving and have some momentum or it would have a very large wavelength and could not be found in the vicinity of the nucleus. Standing waves on a string can be thought of as waves moving in opposite directions and interfering with each other. If we take this approach with an electron, we can think of the wave function having one part moving toward the nucleus and another moving away. The two interfere with each other causing a standing wave. It would look like a pulsating cloud surrounding the nucleus. The pictures I showed you actually represent the electron probability density and not the wave-function itself but it gives you some idea about the wave nature of the electron.

We can still think of the electron as a very small charged object but there is an uncertainty about where this electron is in the cloud which is the size of the atom. If we choose the x direction, we can label this uncertainty Δx. If we try to imagine the simplest spherical standing wave, the diameter of the atom should be on the order of half a wavelength. This means:

$$\Delta x \approx \frac{\lambda}{2} \qquad (8.2)$$

We also know that $\lambda = h / p,$ so:

$$2\Delta x \, p \approx h \tag{8.3}$$

We also have an uncertainty in p. We have no idea which direction the electron is moving. p could be right or left which gives us an uncertainty of $\Delta p_x = 2p$ which results in:

$$\Delta x \, \Delta p_x \approx h \tag{8.4}$$

Heisenberg fine-tuned this idea into one of the most significant principles of quantum mechanics. It's called the Heisenberg uncertainty principle.

$$\Delta x \, \Delta p_x \geq \frac{h}{2\pi} \tag{8.5}$$

The uncertainty in where the electron is located times the uncertainty in its momentum is greater than or equal to h divided by 2π. In other words, if you know where an object is, you don't know what it is doing. The alternative is, if you know what it is doing, you don't know where it is.

The answer to your question is: we can't picture very small things as though they were common objects and we have great difficulty describing objects that are small enough to have wave properties.

J.   Mr. Tweed! I tried to ask a simple question and you gave me the derivation of a major quantum mechanical principle. What would happen if I asked a difficult question? You don't have to answer. Are there any situations where we can actually see this uncertainty principle at work?

Mr. T.   I don't think there are any simple quantum mechanical questions. There are a number of macroscopic things that require quantum mechanics to explain their existence. A permanent magnet is one example. Chemical optical spectroscopy is another. The descriptions of these phenomena all require understanding submicroscopic details which is the realm of quantum mechanics. I can't think of any large scale use for the uncertainty principle. We will, however, encounter it when we study the stars.

K.   Mr. Tweed, you showed us some interesting electron density pictures for l and m not equal to zero. Why did you consider this important information for us to consider, especially since these states are only transitory and exist for only a short time?

Mr. T.   I wanted you to see the similarity between standing waves on a plate (Chladni figures) and electron standing waves calculated from Schrodinger's wave equation. It is not easy to understand why standing waves are so important, but their existence is fundamental for the quantum behavior of matter.

A second reason is that these orbitals become important and are not transient in atoms with many electrons. We will have more to say about this, but first I think it is time for some simulator work

In the simulator

F.   Welcome back; you have been gone for some time.

Today we will spend some time helping you get a feel for atomic sizes. In this simulation you will be able to see photons moving through space. This will be true even for photons outside the visible spectrum. I will set the speed of your clocks to help you view very fast actions.

Let's start by looking at the period at the end of this sentence. You have no difficulty seeing it but if I made it only a little smaller you might miss it. Let's make it bigger until you can see the ink atoms (carbon) grow to roughly the size of the period.

Jill, if you move the wheel in front of you the period will get larger.

J.   I am moving the wheel and the period is the size of the room but I don't see any atoms. I'll make it bigger.

It is now bigger than the size of the room and I still don't see any atoms.

F.   Keep going!

J.   It is now the size of the building and no luck. It is now the size of a football field (~100 *m*) and wait, I think I see them. There are millions of them and they are about the original size of the period.

F.   Why not keep going until you see the nucleus?

J.   OK! I am moving the wheel and now each atom is about the size of the room; no nucleus in sight.

I am turning and turning the wheel and still no nucleus.

There it is, just one small dot. Fritz, you have to help; I have no idea how big the period is now.

F.   It is roughly the size of the earth (~10,000 *km* or 10^7 *m*). Keep going, maybe you will see more.

J.   I am getting a feeling for this wheel. Now the period is about 100 times bigger and I see things within the nucleus. I assume these are protons and neutrons. I'll continue.

Wow, I see three little things inside each proton and neutron that appear about the same size as the original period. Fritz, you have to help again. How big have I made the original period now and what are these things?

F.   It is about 20 times the size of the moon's orbit (10^{10} *m*). The objects you are looking at are called quarks. Mr. Tweed will tell you more about them later.

I hope you can see that matter is nearly all empty space. If a neutral particle such as a neutron is aimed at you, it will not react with the electrons and has a low probability of hitting anything on its way through you.

I have one more exercise before I let you go. I have an atom of hydrogen that I have magnified approximately 2×10^8 times. It now appears as a 1 *cm* ball. Kay, take the atom in the palm of your hand and clap your hands together with the ball in between. Don't clap too hard.

K.   As my hands hit each other the atom jumped in size to about 4 *cm* and then returned to its original size after emitting a flash of light. What happened?

F. You excited the electron from the ground state to *n* =2 and the electron emitted a photon as it returned to the ground state. Clapp really hard this time.

K. Wow! It became very big, almost a meter. It was very difficult to hold onto and it changed into many shapes similar to the ones Mr. Tweed showed us. There was a flash of light in a different direction each time the shape changed and became smaller.

F. Each flash was a photon. The energy from your clap drove the electron into higher quantum states. Kay, can you clap even harder?

K. I will try. I lost it; it just disappeared when I clapped. Wait, I just saw the flashes coming from the other side of the room and now I see it is back to the original size but I lost it. Fritz, please explain.

F. You ionized the atom. That means you gave the electron so much energy from your clap that it has become free of the nucleus. You were then left trying to hold onto a bare nucleus which was so small you could not even see it. I am not surprised that you lost it. Eventually the electron, or another one, was attracted to the nucleus and found its way to the lowest energy level after going through several jumps or possible orbitals.

K. How long does the electron remain in each orbital before moving to the next one or the ground state?

F. I don't know, you have to ask Mr. Tweed and I am not sure he knows. However, it remains in each state long enough to be established (settle) in that energy level. If the electron moved too quickly through the energy levels, the spectral lines would be blurry and not have such exact wavelengths.

Time for you to go. However, I will be seeing you soon.

In the classroom

Mr. T   I hope you had fun playing with the hydrogen atom. We now want to consider larger atoms with more electrons. Let's consider helium. Helium normally has 2 neutrons and 2 protons in its nucleus.   Jill, how many electrons?

J.   The atom must be electrically neutral so it needs 2 negative electrons to balance the 2 positive protons. Do all these electrons go to the lowest energy level (ground state)?

Mr. T.   An Austrian scientist by the name of **Pauli** came up with the conclusion that no two electrons can be in the same quantum state. Another way of saying this is: no two electrons can have the same wave pattern in the same place, they would interfere with each other. It's called the Pauli Exclusion Principle. Therefore, one would think that the two electrons could not be in the ground state. However, Pauli was able to determine that the electron had another quantum property. It's called spin and the electron can have a spin quantum number (*s*) of +1/2 or -1/2.

K.   Why is the quantum number $\pm\frac{1}{2}$ and not some integer like the other ones we have seen?

Mr. T.   It turns out that the quantum number *l* gives us information about the orbital angular momentum of the electron and the quantum number *m* tells us about the value of the angular momentum in a given direction. The angular momentum is quantized in units of $\frac{h}{2\pi}$. The spin angular momentum turns out to come in units half as large as the minimum orbital value (*l = 1*). The formalism was developed before anyone knew about electron spin. Therefore, a quantum number of $s = \pm\frac{1}{2}$ was chosen.

This means that both the electrons in a helium atom can be in states where *n* is 1. One will have *s* = +1/2 and the other has *s* = -1/2.

At this point something should be said about *l* and *m*. *l* is the quantum number for angular momentum and *m* refers to the component of *l* in the *z* direction. The diagram below illustrates this.

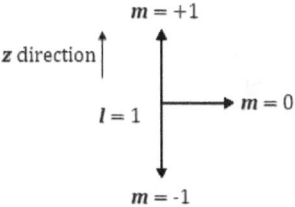

Fig. 8.8

The angular momentum quantum numbers for l =1

The angular momentum is quantized in units of $h/2\pi$. As you can see in figure 8.8, if the angular momentum vector is pointed in the z direction the value of m is +1. If the angular momentum is pointed in the $-z$ direction, m is -1 and in any other direction m is 0.

J.   What is so important about the z direction and how is it determined?

Mr. T. z can be any direction but as soon as it is established, these relationships are established.

J.   It makes no sense to me.

Mr. T.   It's just plain weird. Neils Bohr gave us a quote that I like. *If quantum mechanics has not profoundly shocked you, you have not understood it.* I will have Fritz try to help you with it but he may confuse you even more.

In the simulator

F.   Hi! This little demonstration should give some inkling of how weird quantum particles like electrons are. Kay, take this camera and go into the quantum world room. Jill will be in the room and have some of the properties of a quantum particle. You will not see her; its dark. The only way you can see her is to take a picture.

K.   You are right, it is really dark and I don't see anything. I will take a picture but which way do I point the camera? That's a dumb question. I can't see so even if you told me, I still wouldn't know what to do. I will just shoot. I didn't get anything, the screen is blank.

Here goes again. I got it, I can see her left profile.

F.   Take several more.

K.   This time I see her right profile. I shot again and I see her right profile again.

I keep taking more pictures but all I get is her right or left profiles and it doesn't matter how I hold the camera. I even asked her to turn 90⁰; it makes no difference.

F.   Jill is acting something like an electron. Pick a direction and the electron is either spin up (s = +1/2) or spin down (s = -1/2). Jill shows either a right or left profile.

In the classroom

Mr. T.   The quantum numbers for an electron in an atom are now (n, l, m, & s). l, m, and s are all concerned with angular momentum.

J.   How do we know these electrons have angular momentum?

Mr. T.   Electrons have charge and if charge moves in a circle or spins, it will produce a magnetic field much like a bar magnet. If we place a material in a magnetic field, the energy levels of the electrons will be changed if they have angular momentum.

K.   I don't understand why placing the atoms in a magnetic field will change their energy levels. Mr. Tweed can you give us a simple explanation that does not have a lot of math.

Mr. T.   I will try. If you place a small magnet in a magnetic field it will try to line up with the field. This is why compasses tell us which way is north. If the compass is lined up in the wrong direction it will be unstable and try to seek a lower energy level. If it is lined up correctly, it will do nothing and remain stable.

If an electron in a magnetic field is in an excited state with m = +1 and falls into a state with m = -1, the photon that is emitted will have a slightly different wavelength than if the transition is from m = -1 to +1. We are able to verify the angular momentum calculations by careful study of the spectra.

J.   That seems pretty indirect. Isn't there a more direct way?

Mr. T.   It may be possible to examine the macroscopic magnetic properties of materials.   In most materials the magnetic properties of electrons are cancelled by other electrons. The *B* field is averaged out. However, when there are an odd number of electrons in each atom, it may be possible to see magnetic attraction in large magnetic fields.   Such materials are called paramagnetic.

In ferromagnetic materials interesting quantum mechanical forces tend to line up the fields of the electrons in areas of the structure called domains. In this case we can have permanent magnets and very large macroscopic magnetic effects.

Session 9
Relativity

Since the mathematicians have invaded the theory of relativity, I do not understand it myself anymore.-<u>Albert Einstein</u>

In the classroom

J.   Hi Mr. Tweed! Are we doing more quantum stuff today?

Mr. T.   Only a little bit; in quantum mechanics the effects are important when objects are very small. Today we will look at objects that move very fast (close to the speed of light).

K.   Are we going to learn about Einstein's theory of relativity? I understand it requires the use of very difficult mathematics.

Mr. T.   Yes we are going to take up relativity but we will use simple mathematics and outline only the most important conclusions of the theory.

The starting point for the theory begins before Einstein when scientists wondered about a medium for electromagnetic waves. The idea was: something must be waving; light must be moving through something we can't see that exists even in a vacuum. They called it *ether* and wanted to know if it was stationary or moving.

Late in the nineteenth century **Michelson** invented an incredibly sensitive interferometer that could measure distances with accuracies of a very small fraction of a wavelength of light. This instrument with very few changes is still used in a number of scientific instruments.

Michelson collaborated with **Morley** to use the interferometer to determine if we are moving in the ether. Let's try to understand the interferometer and then examine their experiment.

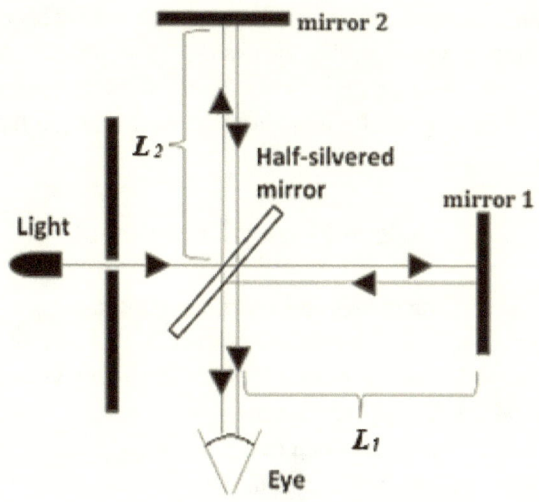

Fig. 9.1
Michelson Interferometer Schematic
Actual unit will contain some optics not shown and in some cases optical detectors replace the observer's eye.

A very thin coating of silver can be deposited on glass that results in part of the light being reflected and part transmitted. Such a mirror is commonly used by police to allow a suspect to be seen but not able to see the observer. In this case most of the light is reflected. In Michelson's case 50% is transmitted and 50% reflected.

In the interferometer, light takes two paths from the light to the observer's eye
1. The light travels from the source straight through the half-silvered mirror, reflects off the mirror 1 on the right, returns and reflects off the half-silvered mirror, and is seen in the observer's eye.
2. The light travels from the source to the half-silvered mirror where it is reflected upward, reflects off the mirror 2 above, returns and passes straight through the half-silvered mirror to the observer's eye.

If everything is adjusted perfectly and L_1 is the same length as L_2, a bright spot will appear in the center of the view. Light from the two paths will add together and reinforcement will occur. If one of the path distances is longer by $\lambda/2$, interference will occur and a dark spot will appear in the center of the view. In some instruments very small distances can be measured by moving one of the mirrors and observing the changes in the interference pattern. In another application one mirror can be replaced by the shiny surface of a specimen. Extremely small vibrations of this surface can then be detected and measured.

Michelson Morley Experiment

Mr. T.   We know the earth rotates on its axis, it rotates in its orbit about the sun, and the sun rotates about the center of the Milky Way. Is the *ether* fixed in space to any of these or does it have a different velocity altogether. Michelson and Morley set out to answer this question. Assume the *ether* stationary and the instrument is moving to the right in the direction of L_1. Let's calculate the times from and to the half-silvered mirror for both paths.

Path 1

The light from the half-silvered mirror starts toward the mirror on the right but as it travels the mirror moves away. The time to reach the mirror takes longer than if the interferometer was stationary. Time to mirror on right = $\dfrac{L_1}{c-v}$ where c is the speed of light and v is the speed of the instrument. The time for the light to come back to the half-silvered mirror will be shorter because the distance the light must travel through the *ether* is reduced. The half-silvered mirror and the light are moving toward each other. The speeds add. Time to return $= \dfrac{L1}{c+v}$. The round trip time (T_1) is the sum of the time out and back.

$$T_1 = \frac{L_1}{c+v} + \frac{L_1}{c-v} \tag{9.1}$$

This can be rewritten as:

$$T_1 = \frac{2 \times L_1}{c} \times \frac{1}{\left(1 - \dfrac{v^2}{c^2}\right)} \tag{9.2}$$

K.   Equation 9.2 looks a lot more complicated than 9.1. Why did you rewrite it in such a complicated form?

Mr. T.   Look at it and tell me what happens when v is close to zero and what happens when v is closer to c.

K. When v is zero the second term in equation 9.2 is 1 and the time is $2xL_1/c$, which is the time with no motion involved. When v is larger, the second term is larger than one, which increases the roundtrip time.

Path 2

Mr. T. When the light travels from the half-silvered mirror upward to mirror 2, it must travel a greater distance than L_1 because mirror 2 has moved to the right by a distance of $v \times T_2/2$ where T_2 is the round trip time. The distance traveled to the mirror 2 is found by using the Pythagorean Theorem.

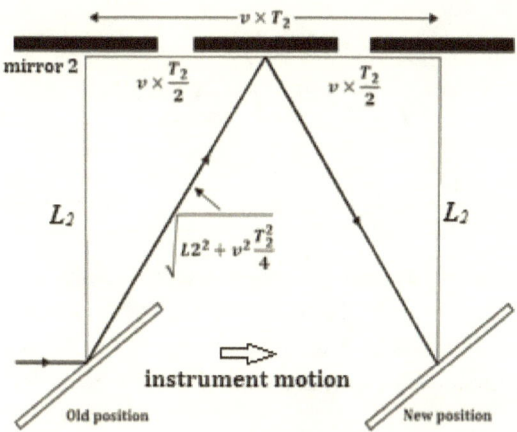

Fig. 9.2

Geometric diagram to find round trip distance from half-silvered mirror to mirror 2

Thus the light must travel a distance of $\sqrt{L_2^2 + v^2 \frac{T_2^2}{4}}$ from the half-silvered mirror to mirror 2 which must equal $cT_2/2$ (the speed of light times the time). This gives us an equation which can be solved for the round trip time (T_2).

$$\sqrt{L_2^2 + v^2 \frac{T_2^2}{4}} = c\frac{T_2}{2} \qquad (9.3)$$

Kay, can you solve it?

K. I will try. I will square both sides of the equation to get rid of the square root sign.

$$L_2^2 + v^2 \frac{T_2^2}{4} = c^2 \frac{T_2^2}{4}$$

Both T_2 should be moved to the same side of the equation.

$$T_2^2 \left(\frac{c^2 - v^2}{4} \right) = L_2^2$$

I think I see what is happening; I am going to try to make it look similar to equation 9.2. I will multiply both sides by $\frac{4}{c^2}$.

$$T_2^2 \left(1 - \frac{v^2}{c^2} \right) = \frac{4L_2^2}{c^2}$$

Now I will isolate the T_2 term and take the square root of both sides.

$$T_2 = \frac{2L_2}{c} \left(\frac{1}{\sqrt{1 - \frac{v^2}{c^2}}} \right) \tag{9.4}$$

It looks similar to equation 9.2 except for the square root term.

Mr. T.   The point is: T_1 and T_2 are different and the Michelson Morley experiment should work. The way they proceeded was to set up the apparatus and then slowly rotate it. This would change which leg was parallel to the instrument motion and a shift in the interference pattern should occur.

Nothing happened! Scientists scrambled to try to explain the result. This was probably the most famous negative experimental result in all of science. One scientist suggested that the dimensions of objects aligned in the direction of motion are shortened.

$$L_1 = L_2 \sqrt{1 - \frac{v^2}{c^2}} \tag{9.5}$$

This shortening of the length (aligned with the motion) was known as the **Lorentz** contraction after the man who proposed it. If you substitute this expression into equation 9.2, you will see that T_1 and T_2 become the same.

A number of years passed until Einstein came up with his Special Theory of Relativity which solved the problem and drove people crazy.

Einstein's hypothesis

All the laws of physics are the same in all inertial frames of reference.

J.   What is an inertial frame of reference?

Mr. If you are stationary or moving in a straight line without any acceleration, you are in an inertial frame of reference.

The hypothesis Einstein made may seem trivial but there are some big consequences.
1. Everyone measures the same speed of light regardless of their motion.
2. If two observers are moving relative to each other, it is not possible to determine if one is stationary and the other moving or vice versa.
3. The time interval you measure depends on how you are moving relative to someone else.
4. Your mass relative to someone else depends on your motion with respect to them.
5. Einstein agrees with Lorentz on his contraction hypothesis.

J.   Is Einstein saying that time is not absolute?

Mr. T.   Yes! We can glean some information about this concept from equation 9.4.

$$T_2 = \frac{2L_2}{c} \left(\frac{1}{\sqrt{1 - \frac{v^2}{c^2}}} \right)$$

The Lorentz contraction only affects distances parallel to the motion. Therefore, L_2 is not affected and $\frac{2L_2}{c}$ represents the roundtrip time (T_0) as measured by an observer moving with the interferometer.   Therefore, equation 9.4 can be written as:

$$T_2 = T_0 \left(\frac{1}{\sqrt{1-\frac{v^2}{c^2}}} \right) \qquad (9.6)$$

This means that T_2 measured by an observer who sees the interferometer moving is longer than the time measured by someone moving with the instrument.

J.   This is crazy! What about the mass? If an object is moving with respect to us is its mass greater or less than before it moved?

Mr. T.   Jill, imagine Kay moving past you at a great speed and she throws you a ball. She throws it at a right angle to her velocity at the precise time so that it will arrive at you. I would not recommend that you try to catch such a ball but let's assume you could, and you could also measure its momentum parallel and perpendicular to her motion. Kay would try to throw the ball at a normal speed but she has a problem; her clock has slowed down. Therefore, her throw will be slow. If that were the only relativistic effect, you would measure the perpendicular momentum as much smaller than she normally throws the ball. You would say, she is the one moving and not me. Einstein says you can't do that. Why not? The answer is that the mass of the ball has increased to compensate for the slow throw.

$$M = M_0 \left(\frac{1}{\sqrt{1-\frac{v^2}{c^2}}} \right) \qquad (9.7)$$

J.   This is even crazier! How do we know if any of this is really true or not?

Mr. T.   We have to look at situations where the velocity v is close to the velocity of light c and I will give you examples but before I do, let's discuss a bit of the General Theory of Relativity.

J.   I hope you are not going to give us any more math. I understand that the math for the general theory was so difficult that even Einstein had trouble with it.

Mr. T.   You are correct about Einstein, some of his teachers thought he was poor in math. It just shows what can be done if you keep working at it. I only have a bit more math. I need some more in order to explain some of the evidence.

General Theory of Relativity

Mr. T.   The special theory did not include gravity or acceleration, either in a straight line or from moving in a curve. The general theory deals with the curvature of space and time and requires tensors to explain its conclusions. We will not deal with curved space or use tensors but we can discuss some of the results of the theory. Here are just a few:

1.   Clocks are slowed not only when something moves fast but also when it is in the presence of a strong gravitational field.
2.   Energy can be directly converted into mass and vice versa.
3.   Gravity acts on photons even though they have no rest mass. When the gravity of a star is so strong that photons can't escape, we have a black hole.

Jill asked how we know if any of this is true. Even though it requires a little math, let's examine one example of time dilation.

Muons are unstable particles, they are like electrons but have much more mass. They are created in the upper atmosphere by high energy cosmic rays. With some fairly sophisticated equipment we can capture some of them and measure their half-life which turns out to be only a little larger than $2 \times 10^{-6}\,s$ ($2\mu s$). They travel at very close to the speed of light. On the top of Mt. Washington more than 150 of these particles strike each square meter per second. The decay of muons can be expressed by:

$$N = N_0 \; 2^{-\left(\frac{t}{2\mu s}\right)} \tag{9.8}$$

Where N_0 is the number of particles at t_0, and $2\,\mu s$ is the half-life of the particles. The elevation of Mt. Washington is 1,917 m, therefore, it takes these particles traveling at close to the speed of light greater than $6 \times 10^{-6}\,s$ ($6\,\mu s$) to travel this distance. Using equation 9.8 we get:

$$N = N_0 \times 0.1$$

This means that only about 0.1 of the number that arrive at the top of Mt. Washington should remain at a detector at sea level in Boston. Guess what, the number hitting Boston is almost as great as Mt. Washington. Their clocks have been slowed down and the muons live much longer than their half-life.

We can spend more time on specific examples where the theory is validated or you can go to the simulator and let Fritz do his thing. What do you want to do?

K.   I have no doubt, let's go to the simulator.

In the Simulator

F.   I am ready for you. You are going to visit a world where light travels at about 50 *km/h*   (~14 *m/s*). This world is set up in a strange way because, if the speed of light were changed, the way electric and magnetic forces act would also change and the world would completely change character. These changes will be neglected and we will concentrate on the relativistic ones.

Let's walk to the train station. We are going on a trip.

J.   I see a lot of flashing lights ahead; it looks like an accident. I hope no one was hurt.

Policeman.   That's the third time this month. Doesn't everyone recognize that this is summer and there are many large insects flying around? You must keep your speed down. You are lucky only the windshield was destroyed and your fender dented.

K.   I don't understand; how did a flying bug cause such an accident?

F.   You remember, the speed of light is only 14 *m/s*. If you are traveling near this speed, the relative speed at which the bug is approaching you is almost the speed of light. It doesn't matter whether you or the bug is moving. If it hits you, you will be struck by a large mass. Remember equation 9.7:

$$M = M_0 \left(\frac{1}{\sqrt{1 - \dfrac{v^2}{c^2}}} \right)$$

As *v* approaches *c*, the term multiplying the rest mass (*M₀*) gets very large. This makes space travel at high speeds very dangerous. If you have watched Star Trek, you will remember they were always putting up their shields to protect the ship from being destroyed by small bits of material in space. Some orbiting satellites have sensors to detect small collisions. They suffer hits on a regular basis but most are small and not traveling at the speed of light.

J. I see the train station. Fritz, where are we going?

F. We are going to California.

K. You can't be serious; California is 36,000 *km* ($3.6 \times 10^7 m$) away and we can't travel faster than the speed of light (14 *m/s*). It will take us $\dfrac{3.6 \times 10^7\ m}{14\frac{m}{s}}$

= 2.6×10^6 *s* which is about a month even if we travel the speed of light. I don't want to stay in the simulator for a month.

F. It won't seem like a month. You have to remember this is a simulator. Trust me.

J. I am not sure we can trust you but I am willing to give it a go. I will buy this large newspaper so that I have something to read on the way.

Train driver. Hello! I have not seen you before. You must be new here. Because you are new, I will explain about the train. Just a minute, here comes my wife and young son. I would like you to meet them. This is Julie and John.

K. I am Kay and this is Jill. We are studying physics and this is a kind of field trip for us. We are trying to learn about relativity. Do you travel with your husband very often?

Julie. Oh, we live on the train, otherwise every time my husband takes a trip my son and I would be a little bit older and after a while my son would be all grown up and I would be old. The company is good to us by allowing us to be together. Now we age at the same rate. I have to go now, it is my job to make sure the train is sealed.

J. We have traveled only a short distance and I hear a loud whooshing sound. What is causing that?

TD. I have moved the train into a section of a large tube which is now being evacuated. When this section has been pumped out, I will open the doors to the main tube which extends all the way to California.

J.   Why does the tube have to be pumped out?

TD. I understand that you saw the result of the car hitting a bug. Our speed will be much closer to the light speed and we want to avoid hitting anything. Even the air in the tube would have a large mass and cause high friction.

There is not much to see in the tube but in order to check our progress, white markers have been placed every 1000 *m*. I will now open the doors to the main tube and we can start.

K.   I can feel the acceleration; I am being pushed into the back of my seat. This is almost one *g* ($9.8\,\frac{m}{s^2}$) of acceleration. It should only take us a few seconds to approach the speed of light (14 *m/s*).

J.   We have been accelerating like this for almost an hour. I wonder how fast we are going. I am going to check the markers the train driver told us about.

This is amazing, each marker is coming every few seconds. We are really moving fast.

K.   We can't move faster than the speed of light but the Lorentz contraction makes the markers close together in our frame of reference. Remember equation 9.5:

$$L = L_0 \sqrt{1 - \frac{v^2}{c^2}}$$

L is the distance we see and *L₀* is the distance between the markers when they built the tube. We are not moving faster but the distance has shortened so that it seems like we are going fast.

The acceleration has stopped. Jill, you can now relax and read your paper.

K.   Here comes the train driver. I am going to ask, how fast are we traveling?

TD. There is no point in giving you a specific speed but we can calculate how close to the speed of light we are moving. The distance between markers appears to be only a few meters, a few hundred times shorter than the original 1 *km*. That means the square root term in the equation you just wrote down must be on the order 1/300. If you have a calculator and do a quick calculation, I think you will find that our speed is nearly 0.99999 times the speed of light.

J.   Our mass must be incredible. What if we hit something like the bug that the car hit near the train station?

TD. That would not be good. If any of the nuclei in the bug's atoms hit a nucleus in the structure of the train there would be a massive shower of subatomic particles.

J.   I think this kind of high speed travel is dangerous and I hope Fritz is taking good care of us.

TD. It's time for you to turn your seats around; we are going to slow down. If you are facing backward, you will be pushed into you seat as we slow down.

J.   I have only been reading for a short while, why are we slowing down now?

TD. We are nearly there and it takes a while to stop.

K.   Wow, I feel the de-acceleration. The distance markers should start to move farther apart now.

TD. California station next everyone. Make sure you have all your belongings and prepare to disembark as soon as this tube section has been filled with air.

Train official.   Right on time, a few minutes ago we received the radio message from the east coast letting us know that you left the station.

TD. Jill and Kay, your ticket calls for an immediate return but you will have a few minutes before we leave. Don't go too far and I will let you know when it is time to leave.

K. Before you go, there is another tube but it is all closed up. Are there other trains to this station?

TD. Yes and one is just coming but doesn't stop here. However you can see it go by.

K. I see it but it is so short. Is it that much smaller than our train.

TD. No, it is the same design as ours. It is just moving fast and looks short from our perspective (frame of reference).

K. Thank you, we will be ready when it is time to go.

J. Look there is a news stand, I am going to buy a California newspaper.

I can't believe the date. It is dated one month after we left. We have lost a whole month, but it seemed like only a couple of hours and the date on my watch is the same as when we left.

TD. All aboard for the east coast.

K. Mr. Train Driver, will the trip back be the same as the trip here? Will we lose another month?

TD. The trip back will be the same but you should not consider that you have lost two months. You will not be two months older. Your friends will be two months older.

K. I can understand why the train driver wants to have his family with him. No one wants to watch their family and friends quickly grow old around them.

F. You have returned from your trip. How was it? How long did it take you to go to California and back?

K. It was a little more than crazy and according to my watch it took only about 4 hours. However the digital clock on the desk indicates that we have been gone two months.

F. When you get back to the classroom, the first thing I want you to do is check the time and date.

In the classroom

K.   I am checking the time and date now. It reads the same date we entered the simulator and only about 4 hours have passed. I guess we can trust Fritz, the whole simulation only took 4 hours.

J.   Mr. Tweed, four hours is a long time. I hope we don't have many more long simulations. I am tired and hungry. Can we stop for today?

Mr. T.   We can stop but I need you to come back in the morning to finish this session. I need to show you some of the material that supports Einstein's theories. We also need to discuss the most important result of all.

The next morning

Mr. T.   I hope you are all rested and ready to go. Should we start with supporting evidence or discuss the important result I mentioned.

J.   I am beginning to know you and I can smell mathematics coming with this important result. We should do that first while we are still fresh.

The equivalence of mass and energy

Mr. T.   Before I attempt to explain this concept, you need to take a quick visit to the simulator.

In the simulator

F.   In front of you is a long (~10 m) open car on a set of frictionless rails. Jill will stand at one end and Kay at the other. There are railings on the ground to hold onto so that you don't start the car rolling as you get on.

K.   We are ready and the car is motionless.

F.   Jill, I am going to give you a heavy ball and I want you to throw it as hard as you can to Kay.

K.   Jill can throw really hard, I am not sure I can catch it.

F.   Trust me, you will catch it.

K.   OK! I guess we decided that you were trustworthy.

J.   I threw it and Kay caught it. As I threw it forward the car began to slowly move backward. When Kay caught the ball the car stopped. The car only moved a short distance.

F.   Kay, can you explain what happened?

K.   The car, Jill, myself, and the ball together represent a closed system. The momentum of the system started out at zero and if there are no outside forces it must remain zero. Therefore, when the ball is in the air, the momentum of Jill, myself, and the car must equal the momentum of the ball. When I catch the ball the momentum of Jill, myself, the ball, and the car are all zero. The only change is the position of the car and the ball is at the other end. We moved some mass from one end of the car to the other.

F.   You got it and we are done.

In the classroom

J.   That was fast, are you sure we got what we needed?

Mr. T.   That was just a little preparation for a thought experiment proposed by Einstein. He loved thought experiments; I am not sure if he ever performed any real ones.

His experiment states: two masses with a total mass M are located at each end of a massless tube of length l. The only difference between the masses is one of them has an electron in an excited state.

The electron drops to its lowest energy state and emits a photon which travels to the other end. The masses and the tube recoil when the photon is emitted and move to the right until the photon is absorbed by the mass at the other end. The second mass now has an excited electron after absorbing the photon.

J. I can see that this is just like our simulator experiment. The two masses are Kay and myself and the photon is the ball. The problem is: the photon has no mass and the ball did.

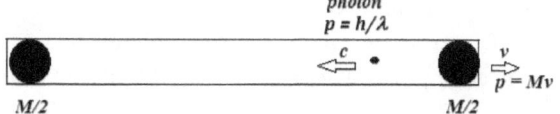

Fig. 9.3

Einstein's thought experiment showing a photon being ejected from the mass on the right

Mr. T. However the photon did have momentum:

$$p = h/\lambda$$

DeBroglie has given us this relationship.

Suppose we think of the photon as having mass m_{photon}.   Then $p = h/\lambda = m_{photon}c$. I used c because the photon travels at the speed of light.

Kay, do you remember the relationship between wave speed, frequency, and wavelength?

K. I do! $\lambda f = c$. It's a photon so we should use c for the velocity. I suppose you want me to substitute λ into the momentum relation.

$$\frac{hf}{c} = m_{photon}c$$

I think I see where you are going. Einstein's photoelectric effect says $E = hf$, so the photon's mass will be:

$$m_{photon} = \frac{E}{c^2} \qquad\qquad (9.9)$$

This is the mass that is transferred from one end to the other.

Mr. T. The idea is we start with an excited electron (one that has some extra energy) at one end of the tube and end up with an excited electron at the other end. There must be some mass transfer because the tube moved, but we only moved energy from one end to the other. **Therefore, energy must have mass**. If we rewrite equation 9.9, we get:

$$E = mc^2 \tag{9.10}$$

This result allows us to calculate the energy of a nuclear reaction. In our sun 4 protons fuse to form Helium nuclei which consist of 2 neutrons and two protons. Two of the protons emit positrons (positive electrons) and lose their charge to become neutrons. Both positrons will find electrons and the two electrons and two positrons will annihilate each other and all that mass will be converted to energy. If you look at the periodic table of elements, a proton has a mass of 1.008 amu (atomic mass unit) and the helium nucleus has a mass of 4.0026 amu. The difference between the mass of 4 protons and 1 He nucleus is 0.0294 amu. 1 amu is 1.67×10^{-27} *kg.* Therefore, the energy produced is $0.0294 \times 1.67 \times 10^{-27} \times (3 \times 10^8)^2 = 4.4 \times 10^{-12}$ *J.* We have neglected the energy of the two electrons that were annihilated.

J.   That doesn't seem like a lot of energy.

Mr. T.   That's the energy of just 4 protons. If you used a gram of hydrogen, the energy would be 9×10^{13} *J* or about 25,000,000 kilo Watt hours. This is more energy than consumed by 2,200 homes in a year.

K.   Do you have other verifications of the theory of relativity?

Mr. T.   There are many but I will limit myself to just two more. Classical physics says the mass of a photon is zero, but Einstein says it has energy, therefore, it has mass and this mass should be acted on by gravity. There are two interesting results involving gravity and light.

Gravitational red shift of spectra from stars

When electrons near a large star jump from orbit to orbit they emit photons with specific frequencies (spectral lines). Gravity acts on these photons and, although gravity can't slow them down, because they travel at the speed of light, they can lose energy as they try to escape the gravitational field and leave the star. Jill, what happens when they lose energy?

J.   Einstein says their energy is *hf*, so their frequency is decreased. The color is shifted toward the red end of the spectrum.

Mr. T.   Yes, this is called a red shift and the general theory of relativity gives the correct change in frequency. What would you expect if the frequency was shifted all the way to zero?

K.   That would require a huge gravitational field. I think we would have a **black hole** where no light could escape.

The bending of light

Mr. T.   If light passes near an object of large mass, the photons will be attracted to the mass and the light will bend. Einstein says this explanation is wrong. His theory predicts that space will have a curvature near the large mass. Light will travel in a straight line but space will curve. This curved space treatment requires math that is beyond our class at the present time, but we can see the effect. It was first seen during an eclipse of the sun. A star that should have been behind the sun was seen at the edge of the eclipsed sun.

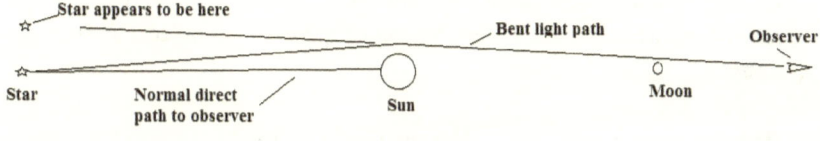

Fig.9.4

Star directly behind Sun seen off to one side during a solar eclipse in 1919

We have a better example. It is called the Einstein cross. Four images of a distant Quasar, believed to be directly behind a large Galaxy, are observed instead of one. The gravity of the Galaxy distorts and bends the light in such a way that multiple images of the Quasar are seen.

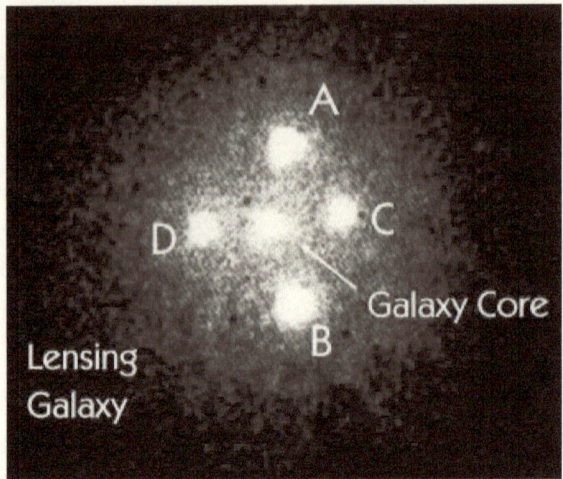

Fig. 9.5

The Einstein Cross

The large mass of the galaxy (probably from a black hole in the center of the galaxy) distorts and bends the light from a distant Quasar in such a way that we see four images of the Quasar. The images are labeled A, B, C, and D.

Quasars produce more radiation than almost any object in the sky. We now know they are not stars. They are called quasi-stellar objects and very far away.

K.   If we have a telescope, can we see the Einstein Cross?

Mr. T.   I don't think so. Quasars emit a large amount of energy as radio waves, infra-red, ultra-violet, and even gamma-rays. I don't know what part of the electromagnetic spectrum was used to produce the image in figure 9.4. It is very possible that it was outside the visible spectrum. If so, it would make seeing the cross with a simple telescope very difficult.

J.   I read somewhere about the clocks in satellites needing relativistic correction. Mr. Tweed, could you explain that?

Mr. T.   The special theory predicts that the satellite clocks will run slower because they are moving. The general theory of relativity predicts they will run faster because they are further from earth. Gravity affects their clock speed. Applications like the global positioning system (GPS) require accurately synchronized clocks. I believe the software that operates these systems compensates for both speed and gravitational effects.

I think it is time to quit for today but before we end, I will leave you something to think about. The train you rode in the simulator moved in a straight line in the tube. What would happen if there was a curve in the tube that forced the train to change direction? Think about it! See you next time.

Session 10

Throwing stuff and seeing the sparks

"Not everything that counts can be counted, and not everything that can be counted counts." (Sign hanging in Einstein's office at Princeton)

J.   Mr. Tweed in session 8 (p. 142) we learned that the nucleus was very small. How do we know that it is so small?

Mr. T.   We throw small things at it.

K.   I don't understand how that will tell us the size; could you explain.

Mr. T.   Perhaps the best way is to go to the simulator and have Fritz help you.

In the simulator

F.   Hi! Good to see you, I was getting bored. I understand that today you will be throwing balls.

K.   Mr. Tweed seems to think that throwing things at the nucleus will tell us how big it is.

F.   He is correct and I have prepared the experiment for you. Do you see the dark fuzzy area in front of you? It is 5 meters high and 5 meters wide and there are 250 objects inside. Jill, you will stand on this side and throw balls into the dark area. Kay, you will go on the other side and catch the balls as they come through.

K.   This is going to be difficult, I don't know where Jill is going to throw and if the ball hits one of the objects, it could bounce in any direction.

F.   I don't think it will be too hard. If a ball is deflected, either catch it or run it down and put it in basket marked *"Hit"*. If it goes straight through, put it in the *"Miss"* basket.

K.   If we throw big balls, we will have a much higher probability of getting a hit.

F.   That's correct. If the object is round, the radius of the target area will increase by the radius of the ball.

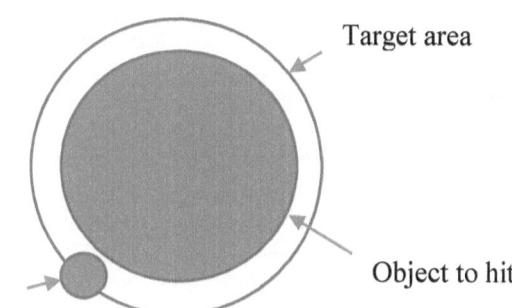

Fig. 10.1

If the center (path) of the ball passes within a radius of the ball, a hit will occur. The target area of each object increases from $\{\pi\, r_{Object}^2\}$ to

$$\{\, TA_{Object} = \pi\big(r_{Object} + r_{Ball}\big)^2\}.$$

There are 250 objects within the cloudy area. Therefore, the total target area is 250 times the target area of each object. Each ball has a radius of 4 *cm*.

J.   Are you ready Kay? I am starting to throw the balls.

K.   The first 4 balls went straight through without any deflection.

J.   I will continue throwing.

K.   Balls number 6 and 8 hit something and were deflected. Ball 11 also was deflected.

J.   Ball 13 came straight back at me.

F.   What does that tell you about the object you hit?

J.   I don't know what you mean.

F.   It tells you something about the mass of the target objects in relation to the mass of the ball you threw. What would happen if the ball had a much larger mass than the object it hit? Would it come back?

J.   I guess not. If you throw a cannon ball at a marble, both will move away from you. Therefore, the mass of the object I hit must be greater than the ball I threw.

F.   Good! Keep throwing balls until we get some good statistics.

J.   I keep throwing and throwing; I am tired. Can we stop now?

K.   I have 35 balls in the *hit* basket and 104 in the one marked *miss*.

F.   I think you have enough information to calculate the radius of the objects in the cloudy area. Kay, give it a try.

K.   I think the ratio of Hits to balls thrown should equal the ratio of the target area of all the objects to the total area. If you add the target area of all the objects and it comes to half the total area, half of the balls thrown should hit objects. This assumes that some of the objects are not hiding behind others.

$$\frac{hits}{hits+misses} = \frac{Number\ of\ objects \times TA_{Object}}{Total\ Area} \qquad (10.1)$$

$$\frac{35}{35 + 104} = \frac{250 \times TA_{Object}}{5 \times 5}$$

The target area of an object comes out to $\frac{35 \times 25}{139 \times 250} = 0.0252\ \boldsymbol{m^2}$

J.   We still don't have the radius of the object but we can calculate the target area radius and subtract the radius of the ball.

$$\pi r_{Target}^2 = 0.0252\ m^2$$

$$r_{Target} = 0.090\ m \qquad\qquad r_{Object} = 0.090 - 0.04 = 0.05\ m$$

The object is only slightly larger than the ball.

F.   Time for class again. I will start preparations for your next meeting.

Back in the classroom

J.   Mr. Tweed our scattering experiment was not too hard but if we wanted to learn about the size of the nucleus, what would we throw and how would we count hits and number thrown?

Mr. T.   Now we can throw many different things but in the 1930's there were only a few possibilities. The choice was to throw alpha α particles.

K.   What is an α particle?

Mr. T.   It is the nucleus of a Helium atom and contains two protons and two neutrons giving it a charge of 2 elementary charges. Helium is ionized in an electric arc and the alpha particles are accelerated and formed into a beam much as we did with electrons in session 4 (p. 54). The first targets were gold nuclei which were fairly large targets. When one of the scattered α particles hits a photographic plate, it records a spot which can be counted. We now have even more sophisticated methods of doing this.

The problem with this method was that the α particle was charged and repelled from the nucleus whenever it was inside the electrons of the target atom. This gave the α particle a larger effective size and it needed significant kinetic energy to overcome the electric repulsion and approach the nucleus. However, **Rutherford** was able to send α particles to within $2.4 \times 10^{-14} \, m$ of the gold nuclei. He concluded that the nuclear radius of gold was probably no larger than $1 \times 10^{-14} m$.

Now we can throw neutrons and they are unaffected by the electrically positive nucleus. However, this property makes them difficult to detect and one has to be clever to form a good beam with the desired neutron energy. High energy neutrons will react in an inelastic way with the nucleus resulting in isotopes and creating other particles in the scattering process. However, neutrons are readily available from nuclear reactors, and neutron scattering experiments have helped us discover much about different nuclei.

J.   How do we know that α particles have only 2 protons and 2 neutrons?

Mr. T.   By studying the electrical and chemical properties of helium we can determine that each atom has 2 electrons. Therefore, the nucleus must have 2 protons. A mass spectrometer might then be used to determine the mass of the nucleus. Chemists have spent years doing many experiments to determine the charge of the nucleus. The periodic table of the elements gives us the atomic number or charge of the nucleus of each element. It was first created in 1871 by **Mendeleev** and is being continually refined.

K. How does a mass spectrometer work?

Mr. T.   First let me explain that the term *mass spectrometer* is a little misleading. The instrument is only able to measure the ratio of mass to charge and not the mass directly. A small amount of the sample to be measured is dissolved into a solvent or made into a fine powder. The sample in placed in the spectrometer and an electric arc ionizes the atoms. This removes the electrons from the nuclei. The nuclei are then accelerated through a known voltage (V) and formed into a beam. (pg. 54 Session 4). The kinetic energy of these nuclei is given by:

$$KE = \frac{mv^2}{2} = qV \qquad (4.2a)$$

The beam of nuclei is then sent into a magnetic field at a right angle to the beam direction. We learned in Session 5 (pg. 71) that a charged particle moving at a right angle to a magnetic field will move in a circle.

$$mv = p = BqR \qquad (5.3)$$

The first of these equations states that the kinetic energy of a charge which has been accelerated by an electric field is equal to the charge of the particle times the voltage that produced the field. The second states that the momentum of a particle moving in a magnetic field can be found by knowing the field strength and the charge (q) and measuring the radius of curvature.

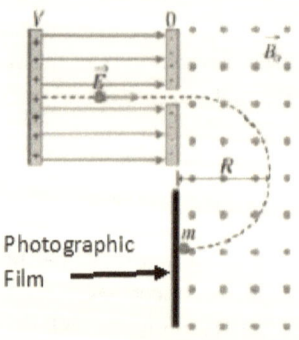

Fig. 10.2

Simplified diagram of a Mass Spectrometer

Kay, what should we do with these two equations?

K.   The only variable in the two equations which we can't easily measure is *v*. Therefore, we should try to eliminate it. You have also said that a mass spectrometer only measures the ratio of **m** to **q**. We should solve for that quantity.

Mr. T.   Kay, go ahead and show us how.

K.   I think there are a couple ways of solving the equations but I would solve Eq. 5.3 for *v* and substitute the result into Eq. 4.2a.

$$v = \frac{BqR}{m}$$, which is substituted into Eq. 4.2a and gives us:

$$\frac{m}{2} \times \left(\frac{BqR}{m}\right)^2 = qV$$, which can be solved for $\left(\frac{m}{q}\right)$.

$$\left(\frac{m}{q}\right) = \frac{B^2\,R^2}{V} \qquad\qquad (10.3)$$

Mr. T.   Good! I think you are ready for the simulator.

In the simulator

J.   Fritz what are we doing today?

F.   We are going to measure the mass of carbon atoms using a mass spectrometer. Our spectrometer will be a little different than the one just described by Mr. Tweed.

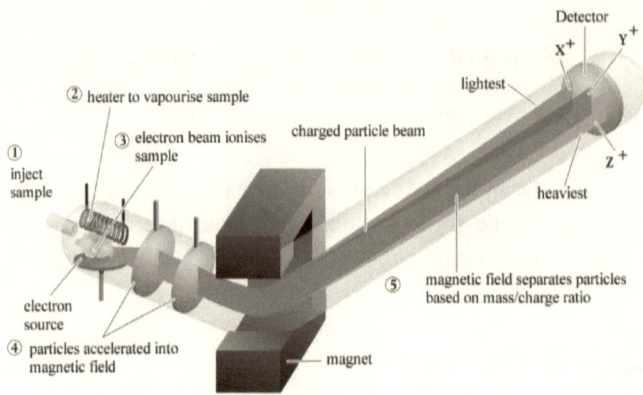

Fig. 10.3

Mass spectrometer with variable magnetic field and fixed radius of curvature. Unlike the photographic film, the detector reads only one mass at a time.

The diagram shows a fixed permanent magnet but it is really an adjustable electromagnet. Jill, I am giving you a small piece of charcoal which we will use for our sample. Grind it into a fine powder and place it the sample container.

J.   I have ground it and placed it the sample container. Now what do I do?

F.   Attach it to the port marked *"inject sample"*. When Kay tells you, you will push the button on the sample container and the carbon powder will be injected into the spectrometer. Kay, turn on the vacuum pump and evacuate the spectrometer. Also turn on the magnetic field and read the field strength value.

K.   The pump is pumping and the magnetic field reads 0.01 *T* (*Tesla*).

F.   We need to wait until the pump has reduced the pressure and there is very little air in the instrument. When Kay says go, Jill will push the button and the vacuum will pull the carbon powder into the spectrometer where it will be ionized and accelerated into a beam of ions. The ionization process has been designed to remove only one electron from each carbon atom giving us an ion with $q = +1$ *elementary charge*. If the magnetic field is too small, the beam will not bend enough to hit the detector. If the magnetic field is too high, the beam will bend too much. The correct radius of curvature for this instrument is: $R = 0.1$ *m*. Kay must increase the magnetic field until she sees a current in the detector. She should then record both the magnetic field and detector current values.

K. The pump has become fairly quiet. I assume this means we have a good vacuum. Jill, are you ready?

J. We're off. The button has been pushed.

K. I am increasing the magnetic field but still no activity from the detector. Now I see some. The magnetic field is 0.110 T and the detector reads 12 units.

F. Increase it some more and see what happens.

K. OK! I see a weaker signal at 0.116 T. The detector current is only 0.011 units. I will try to go higher.
    I am at 0.121 T and I see an extremely weak detector signal of 1.4×10^{-12} units. The detector obviously has a logarithmic scale. I don't understand how a detector can read signals over such a wide range?

F. That is a very good question. The electron multiplier circuitry in one type of detector can read the current produced by a single ion. It is very sensitive. A separate detector would probably be used for the large signals.

K. Why do we have three separate masses? We used pure carbon as a sample.

F. We have found three isotopes of carbon. Each isotope has the same number of protons but a different number of neutrons. The most common isotope is ^{12}C and the next is ^{13}C and the last is a rare isotope ^{14}C produced by cosmic rays in the atmosphere.

K. What does the 12 in ^{12}C stand for?

F. It stands for the total number of nucleons, 6 protons and 6 neutrons. All the isotopes of carbon have 6 protons.

J. We should be able to calculate the mass of the ion from equation 10.3 but you have not told us the voltage (V) and I don't remember the value of an elementary charge.

F. The voltage is 1000 V and the value of an elementary charge is
    $1.6 \times 10^{-19} Coulombs$.

J. Then we can make the calculation:

$$m_{12} = \frac{qB^2R^2}{V} = \frac{1.6 \times 10^{-19} \times .112^2 \times .10^2}{1000} = 2.01 \times 10^{-26} kg$$

K. According to what you just told us, this should be 12 times the value of an
    atomic mass unit (*amu*). When I divide the ion mass by 12, I get
    1.67×10^{-27} kg , which is the value of an *amu*.

J. I calculated the other two results:

$$m_{13} = 2.15 \times 10^{-26} kg = 13\ amu$$
$$m_{14} = 2.34 \times 10^{-26} kg = 14\ amu$$

F. In a real spectrometer the field would be changed automatically and quickly.
    The sample does not last long. The spectrometer output might look like the
    graph of your data below.

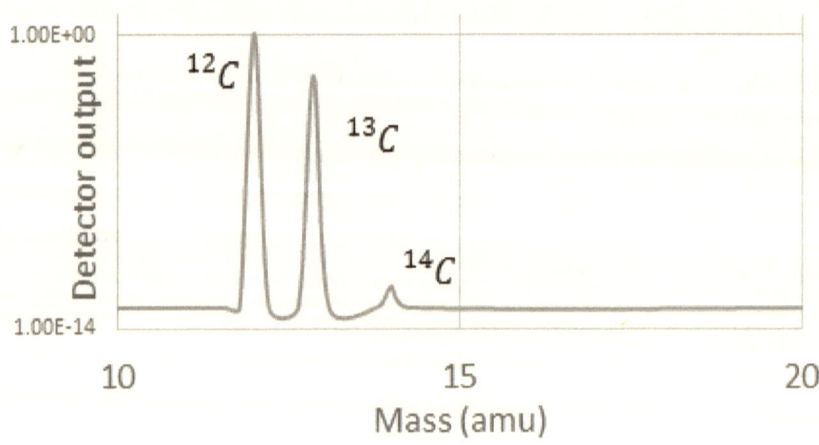

Carbon Mass Spectrograph

Fig. 10.4

Simulated mass spectrograph for carbon.

J. Are there just 3 isotopes of carbon, and what about other elements?

F. Carbon has only 3 natural isotopes; other elements have many more. However, we can artificially create more isotopes of carbon. They are unstable and decay quickly so we don't find them occurring naturally I am going to let Mr. Tweed answer your questions about the other elements. I am going to prepare for your next visit. See you!

In the classroom

Mr. T   I understand that you have been asking questions about the isotopes of the elements. We now have tables that list all the known isotopes as well as their mass, decay mechanisms, abundance, and half-life. Fig. 10.5 also shows the elements and their isotopes. Each little block represents an isotope. A vertical column of blocks shows the isotopes of a particular element. All the isotopes of an element have the same Z or number of protons. As you move up on the chart the number of neutrons changes.

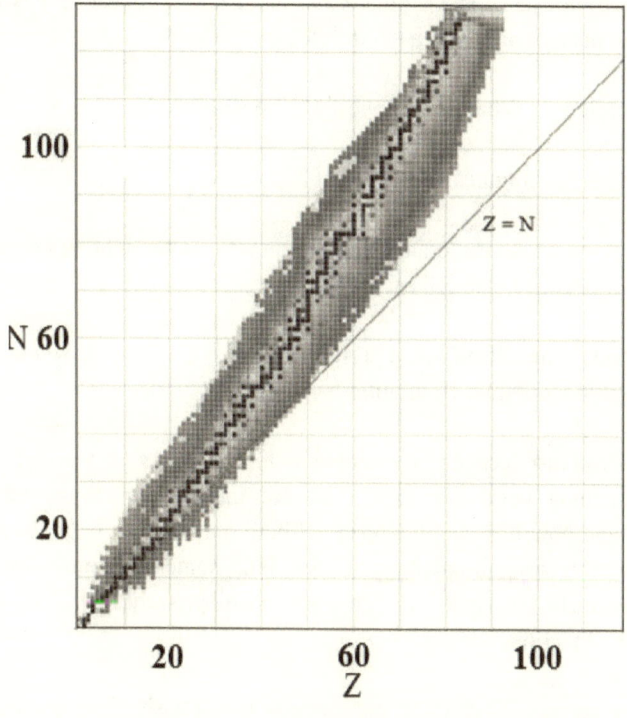

Fig.10.5

Table of the isotopes. Z is a number representing the number of protons and N shows the number of neutrons in a nucleus

J.   What does the jagged dark area in the middle represent?

Mr. T.   The blocks in the middle of the pattern represent stable naturally occurring isotopes.

K.   How are the other isotopes created and how do they decay?

Mr. T.   We throw things at nuclei hard enough that they stick and are either absorbed or break the nuclei into pieces. Both processes can create isotopes. The isotopes decay by emitting electrons, positrons, α particles, or capture electrons.

Maybe Fritz can help you understand a little of these processes.

In the simulator

F.   I am ready for you. The simulator has made nuclei appear very large. I am going to hand each of you a neutron for you to hold.

J.   I am holding it. What am I supposed to do with it?

F.   Just wait a few minutes and see what happens.

J.   Wow! It jumped to the left and an electron went to the right. It's now positive since the negative electron escaped.

F.   It has transformed itself into a proton. Neutrons are unstable when alone and after a little more than 10 minutes half of them will decay.

K.   My neutron just decayed. I was able to hold on to it as it recoiled to the left but something weird happened. The electron also went to the left. I don't understand how this could happen. If both the neutron (now a proton) and the electron had momentum to the left, something must have some momentum to the right. If I throw something away, I will move in the opposite direction.

F.   You are correct. Something did move in the opposite direction but we are not able to see it easily. When scientists first observed this phenomenon, they were very puzzled. They called the unseen particle a **neutrino**. The decay process you have observed is called *beta decay*. The charged particle was called a beta particle; we now know it is an electron. The force involved with beta decay is called the *weak force* and the force holding the nucleus together is called the *strong force*. We have now seen all the forces of nature:

The gravitational force
The electromagnetic force
The weak force
The strong force

Before we do more, I am going to adjust the simulator so that we can see the neutrinos. They move very fast (close to the speed of light) so I am going to speed up your clocks so you can see them move. You will see a bright streak when they are generated.

I have given each of you a proton (^1_1H nucleus) for each hand. I want you to bring your hands together until the protons touch.

K.   It's very difficult to move them together. The closer they come to each other the more force required. I am glad you gave us only a proton. I don't think I could bring together two nuclei with larger electrical charges.

When they touched they grabbed on to each other (*strong force*) and a positive charge was released along with a very fast particle which I assume was a *neutrino* (*weak force*). I could tell that the other particle was positive because it was repelled by the remaining proton; the second proton seems to have lost its charge and changed into a neutron. What was that rejected positive charge?

F.   It was a *positron* which is the antiparticle of an electron. When it meets up with an electron, the two will destroy each other and two powerful *gamma rays* (γ) will be emitted. The remaining particle consisting of a proton and a neutron is called a *deuteron*. Jill, you should also have a *deuteron*.
I am now giving you each another proton which you should push against your deuteron until they touch.

J.   When they touched, their repulsion (*electromagnetic force*) stopped and they became bonded to each other (*strong force*) and I saw a light flash. I assume that was a γ ray.

F.   You now have particles with two protons and one neutron ($^{3}_{2}He$). This is a helium 3 nucleus. Jill and Kay, bring them together.

K.   They bonded together but two of the protons ($^{1}_{1}H$) went flying off in opposite directions. What we are left with is an α particle ($^{4}_{2}He$) with two protons and two neutrons.

F.   This complicated process has changed hydrogen into helium and some neutrinos, positrons, and γ rays. This is how our sun"burns" hydrogen and maintains its brightness. The diagram below may help you see the process.

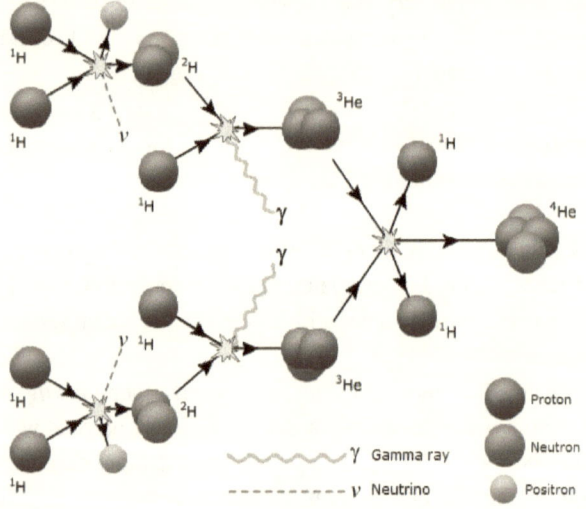

Fig. 10.6

Converting hydrogen to helium by fusion in the sun

J.   What happens to all the neutrinos that are generated in this reaction?

F.   There are a lot of them, but most pass through everything and travel into space. I can give you a demonstration. I will first turn out the lights and simulate just a few coming from the sun so you can get used to the simulation.

J.   I see the light streaks you talked about. Wow! One just passed directly through me.

F.   Remember that your clocks have been speeded up by about a factor of 10^7. This means that when neutrinos from the sun are simulated, you will only see about 1 in 10 million in the room. Here we go.

K.   **Turn it off! Turn it off now!** I am being blinded. I don't see any tracks; the whole room is bright white from the neutrinos. Fritz you should warn us!

F.   I am sorry! There are about 2×10^{14} neutrinos from the sun passing through you every second. They go through everything which makes them extremely difficult to detect. However it is not impossible because there are so many.

K.   I am glad that we can only see them in a simulation. There are too many! I can't even imagine a number as big as 2×10^{14} neutrinos per second.

F.   It is difficult to get a feeling for big numbers. The typical way is to relate the number into some something we can understand. This is not always helpful. For instance: 2×10^{14} birds would be 1000 times the world's bird population. I don't think that helps very much.

These neutrinos pass right through us without any interaction. They are very difficult to detect. Our only hope is the fact that there are so many. Figure 10.7 will give you some idea of the difficulty in making a neutrino detector.

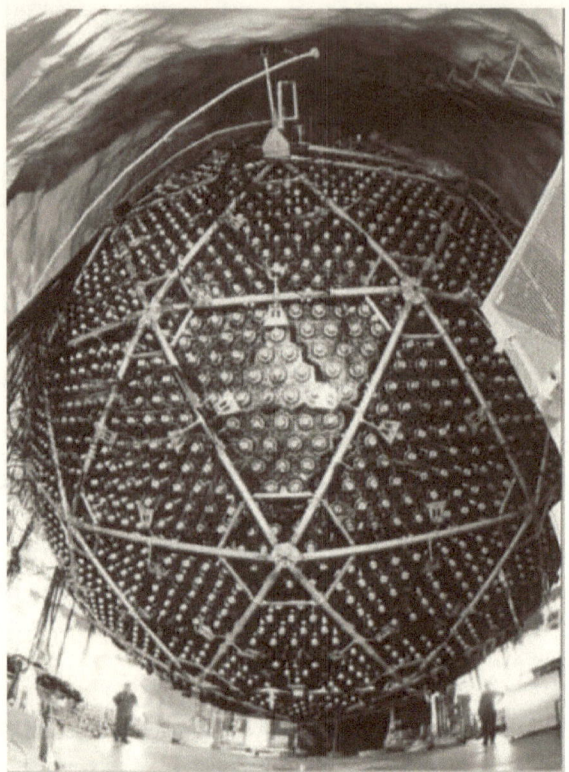

Fig. 10.7

The <u>Sudbury Neutrino Observatory</u>, a 12-meter sphere filled with heavy water surrounded by light detectors located 2000 meters below the ground in Sudbury, Ontario, Canada. Notice the men in the bottom of the photograph.

F.   The heavy water has extra neutrons. Once in a great while a neutron is hit by a neutrino; it is converted into a proton and an electron is emitted. A flash of light is created by the electron as it moves through the water. The light detectors pick up this flash and record it. The apparatus is located below the ground to avoid interference from interaction with cosmic rays and other signals created by man.

J.   I understand that most of the energy generated in the sun comes from the process of converting hydrogen into helium. How much of this energy does the sun radiate in the form of neutrinos?

F. It turns out to be very substantial. It's about 6% of the total.

J. If the neutrinos are so difficult to detect, how do you know it is 6%?

F. We can calculate it using Einstein's famous energy relation that we worked out in session 9.

$$E = mc^2 \qquad\qquad (9.10)$$

The equation gives us the relation between mass and energy. In the fusion process, 4 protons were changed into one helium nucleus, 2 positrons, and 2 neutrinos. Very careful mass measurements with mass spectrometers and other instruments give us very accurate measures of the mass of these particles.

proton mass	1.007825 *amu*	1.683x10^{-27} *kg*
electron mass	5.45x10^{-4} *amu*	9.10x10^{-31} *kg*
positron mass	"	"
neutrino mass	unknown	very small
$_2^4He$ or α particle	4.002603 *amu*	6.6843x10^{-27} *kg*
1 *amu* = 1.67x10^{-27} *kg*		

Kay, add up the mass in *amu* before the fusion process. Jill, find the mass after the fusion process.

K. Before is easy, it's just the mass of 4 protons or 4.0313 *amu*.

J. I have to add the masses for one α particle, two positrons, and two neutrinos.
$$4.002603 + 2\text{x}5.45\text{x}10^{-4} + 0 = 4.0037 \text{ } amu.$$

K. Some mass was lost (4.0313-4.0037 = 0.0276 *amu*). Where did it go?

F. Some went into heat (kinetic energy of particles), some went into radiation, and some went into neutrino production.

J. How many joules of energy is 0.0276 *amu*?

F. If you convert it into kg you can use Einstein's relation (equation 9.10). Remember, the speed of light c is 3x10^8 *m/s*

J. I will try: $0.0276 \text{ } amu \times \dfrac{1.67\times10^{-27} \text{ } kg}{1 \text{ } amu} = 4.609\text{x}10^{-29} \text{ } kg$

$$E = mc^2 = 4.609 \times 10^{-29} \times (3 \times 10^8)^2 = 4.14 \times 10^{-12} J$$

K. That doesn't seem like very much energy.

F. You have to remember that this is the energy for burning just 4 protons. Can you imagine how many protons are in the sun? The sun is mostly hydrogen. Sometimes this energy is expressed in the units of electron volts (eV).

$$4.14 \times 10^{-12} J = 2.59 \times 10^7 \ eV$$

Which is almost 26 million electron volts. It's the same amount of energy but it seems greater doesn't it?

J. Can the sun burn nuclei other than hydrogen.

F. Nuclei with more protons repel each other with a greater force and must approach each other with high energy in order to touch and fuse. In session 2 we discussed the fact that the average kinetic energy of gas particles determined their temperature.

$$\frac{2\overline{KE}}{3} = k_B T \tag{2.10}$$

When scientists first calculated the kinetic energy of protons in the sun, they concluded the temperatures were too low to create fusion. This was clearly wrong. The problem was finally solved by including the wave (quantum mechanical) nature of the protons in the calculations which allowed a very small percent to fuse - just enough, in fact, to keep us warm. I think you can begin to appreciate why it is so difficult for us to develop a fusion nuclear reactor for power generation.

However, the sun is not hot enough to fuse heavier nuclei in any significant numbers. Some very hot stars will fuse more nuclei and produce heavier elements. Common stars will, however, not produce elements heavier than iron. This is why the earth is composed mostly of iron and nickel. We gain energy from fusing light elements. Fusing large nuclear masses requires energy and does not create it. In later sessions we will talk about where the heavy elements come from.

J. We get nuclear energy from fission in our reactors. Does the sun get any energy from fission?

F.   We don't think it does. We have no spectroscopic evidence that there is a significant amount of heavy elements in the sun and there is no reason to believe that there are enough neutrons to sustain a chain reaction.

J.   Explain a chain reaction.

F.   When a nucleus like $^{235}_{92}U$ is struck by a slow neutron which is absorbed, it becomes excited and splits into two pieces and some more neutrons. If the sample is small, most of the neutrons escape. However, if the sample is large, some of the neutrons will split other nuclei and they will emit even more neutrons. We get a chain reaction. If we don't control it, we get a bomb.

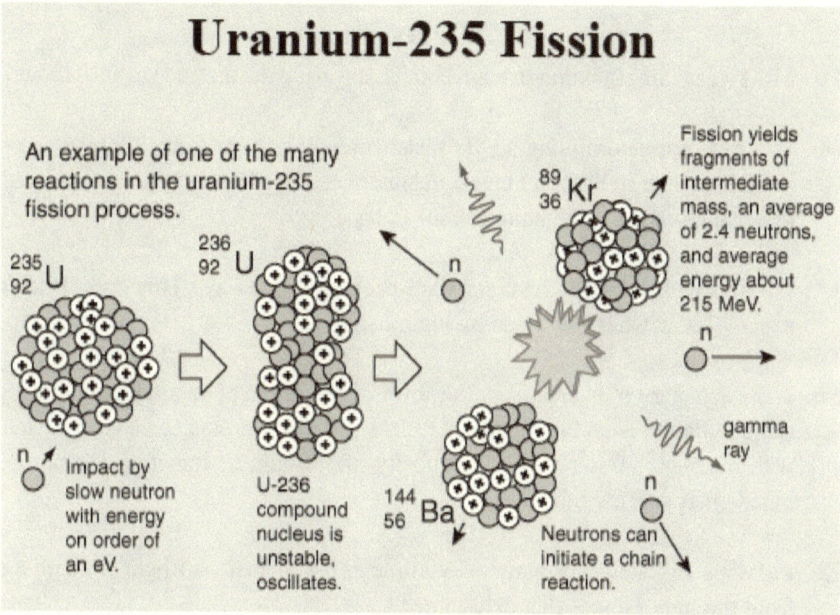

Fig. 10.8
Nuclear fission

J.   That is all very complicated. What happens when our sun burns up all its hydrogen?

F.   I am going to leave those questions for Mr. Tweed in the next session. That's all for today. See you next time.

Session 11
How far is far?

No conception whatever can be had of the magnitude of the visible universe until the distances of the stars are known. None of the millions of human beings that have (sic) lived and dies knew the distance of even one star from the earth until within the last seventy years. . . . The word millions has for long been used in telling the number of stars. But billions now appears to be more appropriate. Each one is a hot sun, and each may be attended in many cases by inhabited worlds.— <u>Edgar L. Larkin, 'Measuring the Distance of a Star,' Scientific American, 28 October 1905</u>

In the classroom

J.   Mr. Tweed, are we going to learn about measuring the distance to stars today?

Mr. T.   Yes, but determining stellar distances will require some knowledge of luminosity and stellar brightness, and understanding how brightness is related to distance. That's the main task for today.

J.   I understand that very faint stars are probably far away. However, I don't exactly know what you mean by luminosity.

Mr. T.   Each star emits energy in the form of radiation. The total amount of a star's radiation is its luminosity (L). It is usually expressed as a comparison with our Sun. A star with a luminosity of 5 would be radiating 5 times as much energy as the Sun.

K.   I assume by "stellar brightness" you mean the intensity of light (I) coming from the star. How is that determined?

Mr. T.   You have to measure the intensity of the light (energy per square meter) coming from the star. We now have special instruments that will measure how much light is striking the image of the star in a telescope. These are similar to the solid state image devices found in your cell phone or digital camera.

In the past the astronomer would use a scale of brightness developed by the Greek Hipparchus to compare stars. The brightest star in the heavens was given a magnitude 1. The weakest that could be seen with the eye was a magnitude 6. This created a scale of 1 to 6 which was quite useful in comparing stars. However, not only did this require considerable experience on the part of the observer, but some complications were introduced because the response of the eye is not linear. A magnitude 1 star is 100 times brighter than magnitude 6. A more modern version of this system is in use today. In fact today's newspaper indicated that the international space station could be seen without a telescope and its magnitude was about -3.7. That's very bright.

When a star shines, its light spreads out in a sphere that increases its area as the square of its radius. This means the light intensity (energy per square meter I) is inversely proportional to the distance (D) squared. It is also directly proportional to the luminosity (L).

We can combine these two relations to form one simple proportionality statement:

$$I \propto \frac{L}{D^2} \qquad\qquad (11.1)$$

K.   This proportionality statement can be valuable for comparing two stars but I will not be able to determine the distance to one of them unless I know the distance to the other. We must make some direct distance measurements.

Mr. T.   Correct! We are going to use parallax measurements to determine the distance to the nearest stars. Fritz will explain.

Parallax

In the simulator

F.   You use parallax all the time to determine the distance to close objects. Each eye sees a different image from the other one. A close object will shift its position in the field of view from the left to the right eye because your eyes are separated. This shift allows you to determine the distance. It happens without you being aware of making any mental calculations; your brain just does it automatically. Today we will try to make you aware of what your brain is doing. I want you, first, to determine two things. 1. What is the distance between your eyes? 2. What is your field of view in degrees?

K.   It is easy to measure the distance between our eyes (0.07 m), but I don't understand what "field of view" means.

F.   If you close one eye and stare straight ahead, you can still see things on the left and right. The angle between what you can just see on the left to what you see on the right is your field of view.

J.   We can help each other. I will stand facing the wall and stare at a spot with one eye. Kay can move her hand from the far left until I just see it and then she will place a piece of masking tape on the wall. She can repeat the process for the right side. We now have an angle from the left tape to my eye to the right tape, but I don't know how to measure the angle.

K.   We can measure your distance from the wall and the distance between the tapes. The angle can be calculated using trigonometry.

J.   The distance to the wall was 1 *m*
     The distance between tapes was 1.1 *m*

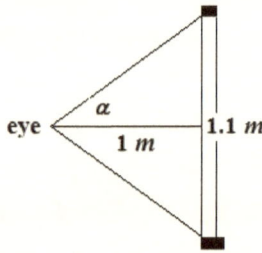

K.   I can calculate the angle α. We need a right triangle, so I will first calculate $\alpha/2$. $\frac{\alpha}{2} = tan^{-1}\left(\frac{0.55}{1}\right) = 29° \text{ or } \alpha = 58°$

F.   We can use 60°. I will give you a separate view from both eyes and you can see the star shift position in the field of view.

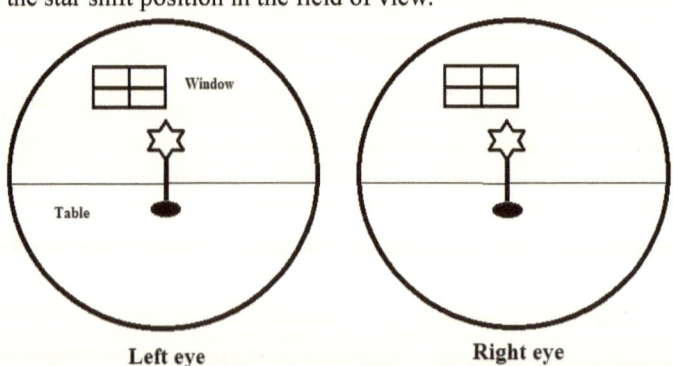

Fig. 11.1

Paper drawing of right and left field of view images of room a near object on a table

K. I can see the star has a different position with respect to the window. I assume we are going to figure out what angle is involved in the shift and draw some triangles to determine the distance to the star. I will first make a paper drawing of the fields of view. How are we going to figure out how many degrees the star shifted with respect to the window?

J. I can figure that out. We determined the field of view was 60° so the diameter of your view drawing represents 60°. We have to measure the field diameter and star shift in *cm* from the drawing and use proportion to determine the angle of shift (α).

K. The diameter of the field is 8.25 *cm* and the star only shifted 0.3 *cm*.
60° is to 8.25 *cm* as α is to 0.3 *cm.*

$$\frac{60}{8.25} = \frac{\alpha}{0.3} \text{ or } \alpha = 2.2^\circ$$

I will try to draw the triangle.

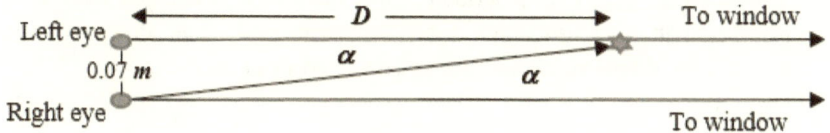

Fig. 11.2
Diagram showing the triangle needed to calculate the distance to the star

The triangle consists of the unknown distance **D**, the distance between our eyes, and the angle α.

$$\tan 2.2^\circ = \frac{0.07}{D}, \text{ which gives } D = 1.8 \ m$$

F. Your eyes are pretty good at parallax distance sensing out to a few meters, and your brain takes care of all the math automatically. The *base line* for these estimations is the distance between your eyes. Larger distances require larger baselines. The largest *base line* we have is the diameter of the earth's orbit ($\sim 3 \times 10^{11} \ m$). The usual procedure is to observe a star many times over a period of several years. If the image of the star shifts significantly against the background of faraway stars, it must be close. Mr. Tweed may have more on this.

In the classroom

Mr. T.   In 1989 a satellite named *Hipparcos* was launched to survey the close stars in the Milky Way. It was able to gather data on more than 100,000 stars and determine their distance using parallax.   The name stands for **Hi**gh **P**recision **Par**allax **Co**llecting **S**atellite. Before satellites such as *Tycho* and *Hipparcos,* parallax was valuable for distances up to 400 light years (**LY**). I don't know what the new limit is but it should be much larger.

J.   Hipparcos must have a very powerful telescope, the parallax angles must be very small.

Mr. T.   The telescope is very strong but its light gathering power need not be so great because it is only looking at stars that are close. The big advantage of a satellite is there is no air pollution and thermal distortion.

Data for one of the stars in the Hipparcos catalog is shown in figure 11.3. The data was taken over a period of more than three years. You can see the parallax pattern repeat, one loop for each year. The dots represent measurements and the curved line is calculated from the data.

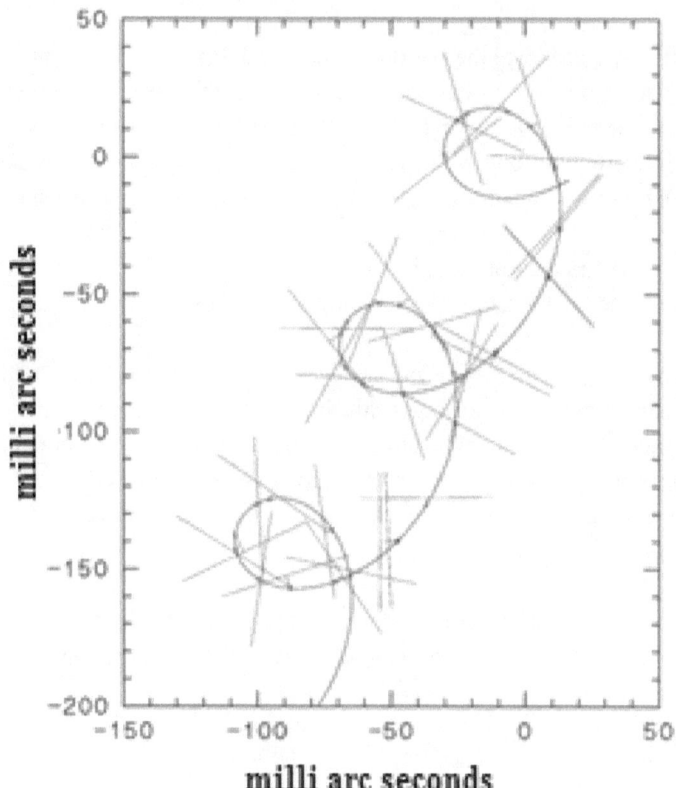

Fig. 11.3
Observations by the satellite Hipparcos of the parallax and motion of a star

K. I can understand why the star image should shift back and forth as the earth travels around the sun, but why is it moving down the page?

F. This is a very interesting case because the star is moving and Hipparcos is not only able to determine the distance but also the motion of a star.

J. Kay, how did you know the star was moving down and not up the page?

K. I assumed that the first measurements would be at an angle of zero and as time went on, the vertical and horizontal angles both decreased.   The measurements at the bottom of the page were taken later than at the top.

Mr. T.   Do you think we can estimate the distance to the star from this diagram?

J.   When we considered the star on the table with Fritz, we determined the angle shift from one eye to the other and knowing the baseline we were able to determine the distance. In this case the baseline is the diameter of the earth's orbit (3×10^{11} *m*) and we need to determine the angular shift caused by the earth's motion. We must neglect the star motion but why is the star moving?

Mr. T.   The stars are not fixed in the sky. They rotate around the center of our galaxy and have other somewhat random motions.

K.   After mentally subtracting the steady motion of the star, it looks to me like the earth's motion is causing a shift in the angular position of about 50 milli arc seconds (*mas*). I am just looking at the size of the loops but I am sure the people analyzing the data from Hipparcos could make a more accurate determination. However, we should be able to make an estimate using my number.

J.   We need to change 50 *mas* to degrees before my calculator will work.
I will convert to degrees and complete the distance calculation.

$$50 \; mas \times \frac{1 \; as}{1000 \; mas} \times \frac{1 \; deg}{3600 \; as} = 1.4 \times 10^{-5} \; deg$$

We can use the same math as we did with Fritz.

$$\tan \alpha = \frac{baseline}{Distance} \qquad \tan(1.4 \times 10^{-5}) = \frac{3\times10^{11} m}{D}$$

$$D = 1.2 \times 10^{18} \; m$$

K.   That's a very large number. Don't astronomers use light years as the unit of distance.

F.   The speed of light is 3×10^8 *m/s* and there are about 3.16×10^7 *s/yr*. Figure out how many meters are in a light year.

J.   That's easy. The distance is the speed times the time or:

$$3 \times 10^8 \times 3.16 \times 10^7 = 9.5 \times 10^{15} \; \frac{m}{LY}$$

K.   Now we can figure out how far the star is in light years.

$$D = 1.2 \times 10^{18} m \times \frac{1 \; LY}{9.5 \times 10^{15} m} = 126 \; LY$$

Mr. T.   You can see that the loops will be small for a star as far away as hundreds of light years. We need to understand how we determine distances much greater than this. The first part of this involves understanding the relationship between color, temperature, and luminosity. Fritz is waiting for you.

Color and luminosity
In the simulator

F.   I have something hot for you to look at. I will turn out the lights and see if you can see it.

J.   I don't see anything, it's very dark in here.

F.   Let me heat it to a higher temperature. Can you see it now?

K.   I can just barely see a dim dark red rod.

F.   Let me heat it up to about 2000° *K*.

J.   I can see it clearly now, it's a nice red.

F.   I am going to go all the way up to 5000° *K*. What does it look like now?

K.   The whole room is brightly lit up and it is much too bright to look at. The light is white.

F.   In the real world I couldn't do this without the rod melting but in the simulator I can increase the temperature of the rod to $10,000^{\circ}$ *K*.

J.   Everything is blue and so bright I can't stand it. Please put it away.

F.   The luminosity (total power of the light emitted) has a more complicated relationship but you could clearly see that high temperatures yield more luminescent objects. I am sure that Mr. Tweed will be able to help you understand how all this fits together.

In the classroom

Mr. T. Kate stated that we would have to make some direct distance measurements in order to compare stars using the proportionality statement 11.1 ($I \propto \dfrac{L}{D^2}$). The method of parallax gives us a way to do this. Once we have measured the distance (D) and light intensity (I) we can compare the luminosity (L) with our Sun. We can, therefore, produce a large table of all the close stars. I will give you some data to play with:

Star	I (rel. to Vega)	T (deg. K)	Luminosity	Dist (LY)
Vega	1	9,600	50.1	25.4
Tau Ceti	.0409	5300	0.44	11.9
Sun	5.15×10^{10}	5800	1	1.58×10^{-5}

Table. 11.1
Data from three stars showing the light intensity (I), temperature (T), luminosity (L), and distance.

Kay, see if you can calculate the luminosity of Tau Ceti from the Vega data.

K. The luminosity is already given in the table, I don't need to calculate it.

Mr. T. I understand that. Just pretend you don't know and do the calculation to see if you get the same number that's printed in the table.

K. I can try. I will first convert the information in relation 11.1 into an equation involving both stars. I will call Vega star 1 and Tau Ceti star 2.

$$I \propto \frac{L}{D^2} \quad \text{converts to:} \quad \frac{I_1}{I_2} = \frac{L_1}{L_2} \times \frac{D_2^2}{D_1^2} \text{ which can be solved for } L_2$$

$$L_2 = L_1 \times \frac{I_2}{I_1} \times \frac{D_2^2}{D_1^2} \qquad L_2 = 50.1 \times \frac{0.0409}{1} \times \frac{11.9^2}{25.4^2}$$

My calculator tells me L_2 is 0.450. What is wrong? The table said it was 0.44.

Mr. T. You did nothing wrong. Your value was very close.

K.   Why are we assigning a value of 1 to the light intensity of Vega and comparing other stars to it?

Mr. T.   Vega has a value of $m = 0$ on the scale devised by **Hipparcus**. Hipparcus originally defined $m = 1$ as the brightest star in the sky. I don't know how he missed Vega but it is one unit of magnitude brighter than the stars he classified as brightest. One of the brightest objects we see in the sky is Venus and it has a magnitude value of about -4 which is very bright.

J.   Mr. Tweed you said previously that the satellite Hipparcos made measurements on more than 100,000 stars. How does anyone make sense out of such a huge amount of data?

Mr. T.   The most famous diagram in astronomy is the Hertzprung-Russell diagram. This diagram is a plot of luminosity against the surface temperature of stars ranging from high temperature blue-white stars on the left side of the diagram to the low temperature red stars on the right side.

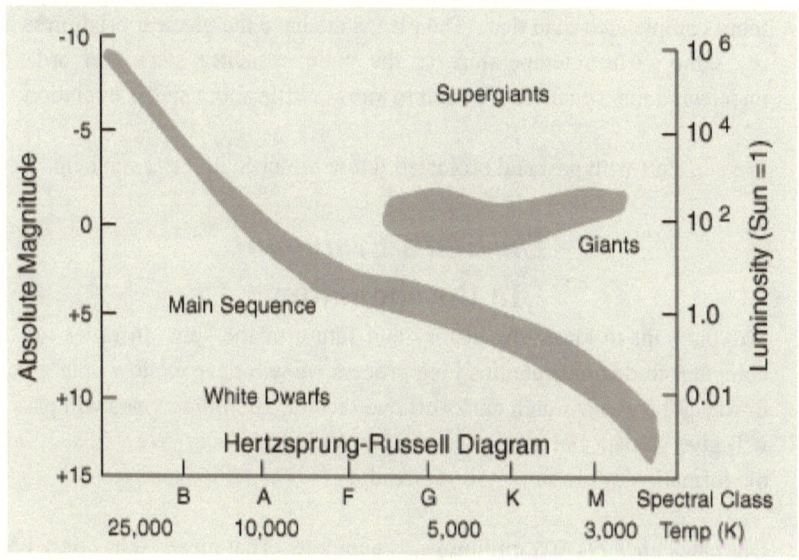

Fig. 11.4

A Hertzsprung–Russell diagram relating the luminosity and surface temperature of many stars

Our sun is a member of the main sequence of stars in the Hertzsprung-Russell (HR) diagram shown in figure 11.4. You can see approximately where it lies by knowing its luminosity has a value of 1. The luminosity scale on this chart compares each star to the sun. Thus, a star with a luminosity of 10^2 will produce 100 times as much light as the sun.

I will summarize the main ideas shown in the diagram:

1.   Most stars fall in a band called the *Main Sequence.*
2.   The letters at the bottom refer to optical filter designations that are used to determine the color of the star.
3.   Hot stars are blue (short wavelengths) and highly luminescent.
4.   Cool stars are red (long wavelengths) and relatively dim.
5.   The luminosity scale uses a value of 1 for our sun. Thus the most luminous stars produce 10^6 times as much light as our sun.

If we want to determine the distance to a star, the first step is to determine its luminosity. If we only had stars in the main sequence it would be relatively easy. We could measure their color which would give the temperature and then look up the luminosity from the HR diagram. Unfortunately, it's a little more complicated than that. The giant stars have the greatest brightness but the same surface temperature as the main sequence stars. In order to understand this situation, we need to know a little about stellar evolution.

We can start with past and projected future of our Sun. Fritz can help.

The Sun's Evolution
In the simulator

F.   So you want to know the history and future of the sun. In order for the computer to demonstrate this long process we will have to slow your clocks down. When your watch clicks off one second, 10 million years will pass. I will give you special glasses so you can look at the sun. We will start with the formation of the sun. Are you ready?

K.   Our clock speed is 600 million years a minute. That means it has only been about 6 seconds since the dinosaurs went extinct. I guess I understand; let's go.

J.   It's dark, I can see the stars but it looks a little hazy.

K.   Now it's really hazy. I can no longer see the stars.

J.   I can see a big red luminous ball that takes up nearly the whole sky. Fritz, tell us what is happening?

F.   Gravity is pulling the stellar dust together and it is getting hot.

K.   Wow, suddenly it got light and the Sun is shining.

F.   The temperature in the center of the Sun rose to about 1 million degrees and that started the fusion of Hydrogen.

J.   We started this process a little over 7 minutes ago and the Sun looks normal although just a little brighter than when it first lit up.

F.   It looks normal because the time is the present. We will now be simulating the future.

J.   It's been about 12 minutes and the sun looks about the same but I am getting warm.

F.   I will turn on the air conditioning and provide you protection against the radiation.

K.   It's been 16 minutes and I can feel the increase of the sun's radiation even though you have provided protection. Fritz, what is happening?

F.   The sun has just about burned all the Hydrogen in its core.

K.   If it has burned up its fuel, why is it getting hotter?

F.   It's only burned up the Hydrogen in its core and now the core is shrinking because there is no energy being produced to counteract the gravity. As it shrinks, it heats up.

J.   Look the sun is getting larger fast, and even though we are protected I can tell it's getting very hot. What's happening now?

F.   The Hydrogen in the core is gone but there is more in the outer shell of the star. Fusion of Hydrogen is now happening outside the core. The rapid expansion is cooling the outer surface and the sun is now red.

K. Only 17 minutes have passed and the sun is huge and takes up almost a third of the sky. The heat must be incredible. I am glad we are protected.

F. The Helium in the core is now burning as well as the Hydrogen just outside the core. The sun is so large that the planet Mercury has been absorbed. The luminescence is almost a thousand times greater than when we started. It is now a *red giant.*

J. 18 minutes have passed and the sun continues to grow and get hotter.

F. The core now consists of Nitrogen and Oxygen which came from the fusion of Helium. Both Helium and Hydrogen are in short supply in the core. Hydrogen still exists in the outer shell.

K. 19 minutes have passed and the outer layer seems to be unstable and oscillating in size. The sun takes up nearly the whole sky. Look the outer layer just got blown away from the center. But there is still a small hot core.

F. The outside that has been blown away is called a planetary nebula.

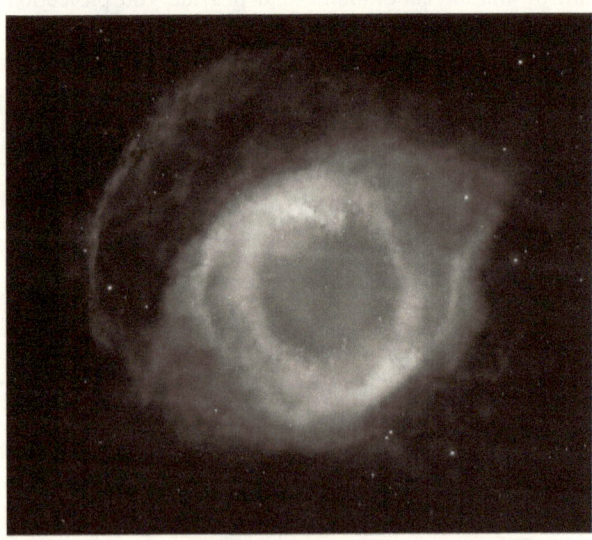

Fig. 11.5
The Helix Nebula
A planetary nebula picture from NASA taken with the Hubble space telescope.

J. Did our Sun explode? Look, there is a tiny bright star where our sun was located.

F. No, it just rapidly pushed away its outer layers. The center core became a white dwarf star. No energy is being produced in this star but it is so small and dense that it will take a long time to cool. When it cools off and becomes a black dwarf, it will not be a great deal larger than Jupiter. The following diagram will summarize what you just experienced in the simulator.

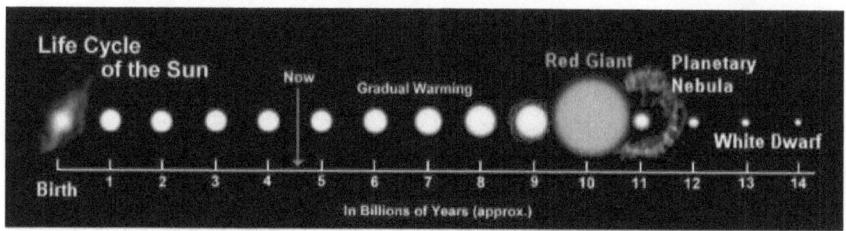

Fig. 11.6
Time scale for the evolution of our Sun
Maybe Mr. Tweed can tell you about other stars. See you!

The Evolution of Large Stars

In the classroom

Mr. T. The larger stars have a slightly different history than our Sun.

K. Our Sun started out on the main sequence and moved up into the giant part of the HR diagram. Will the large stars do the same?

Mr. T. Our model of stellar evolution has all the stars beginning their lives on the main sequence where they are burning (fusing) Hydrogen. The larger stars start further to the left and have higher surface temperatures and are more luminous.

J. Will these large stars live longer than our Sun?

Mr. T. Actually, the large stars live shorter lifetimes. Let's look at the example of Bellatrix.

	T (deg. K)	L	Mass	Size	Dist. (LY)
Bellatrix	22,000	6,400	8.4	6	250
Sun	5,800	1	1	1	0

Jill, how much faster is Bellatrix losing mass than the sun?

J.   It that a trick question? The answer is 6,400 times. The mass loss is from radiation of photons and their mass can be calculated from their energy (hf) using Einstein's formula $E = mc^2$. The luminescence is the total rate of energy loss from radiation.

K.   I see it. It has 8.4 times the mass of the Sun but is losing it 6,400 times as fast. It's not going to have a long life.

Mr. T.   We will not do the math here but Bellatrix is estimated to be no more than 20 million years old and it has nearly burned all the hydrogen in its core. It is now leaving the main sequence.

J.   It's so young! 20 million years is only two seconds of Fritz's simulator time.

K.   Will it become a red giant like our Sun did in the simulator?

Mr. T.   Eventually it will become a red super-giant but it will probably go through some interesting stages before that. It will grow into a blue supergiant and burn helium. It will then go through stages where it burns (fuses) carbon, oxygen, and finally silicon. The fusion of these larger nuclei will create elements with larger nuclear masses. The largest of these nuclei will be Iron. It will change its color to Yellow and finally become a red super-giant. Its luminosity will not change very much during this time; it moves almost horizontally across the Hertzsprung-Russell diagram.

J.   Why will larger nuclei not form?

Mr. T.   The big nuclei have a large electrical charge and it is very difficult for other nuclei to approach them. The positively charged approaching nuclei will be repulsed unless they have very high kinetic energy (temperature). The temperature required is greater than even the extreme temperature at the center of this star.

At the end it will most likely become a Type Ia supernova (explosion) and its central core will become a black hole.

J.   What is a black hole?

K.   It is an object so small and with so much mass that light cannot escape its gravitational field. Light that comes closer than the *event horizon* will be sucked in and disappear forever.

J.   That means they don't radiate and light will not reflect from them. How do we know they even exist?

Mr. T.   We look for gravitational effects at fairly large distances. A star may rotate around a black hole. You couldn't see the black hole but you could see the star in its orbit around the hole.

However, before a large star like Bellatrix blows up it might do something very interesting and valuable to us. At some point the atomic gas in the outer layers will be mostly Helium. If it is ionized it will be relatively opaque and absorb more of the radiation from the star. This will heat it up and it will expand. As it expands it will cool, collect electrons, become less ionized, and become more transparent. Without the extra heat from absorbing radiation the helium is pulled closer to the star, becomes ionized, and the whole process repeats with typical periods of days to years. This means the star will pulsate in a regular fashion. When the Helium is ionized, it will appear dim; when the gas is more transparent, it will be brighter. It will become what is known as a Cepheid variable.

K.   Why did you say this would be valuable to us?

Mr. T.   If we wish to measure distances to distant stars, the first step is to determine the star's luminosity. It was discovered that the pulsation period of a Cepheid variable is directly related to its luminosity. If we can measure the pulsation rate, we can calculate the luminosity even if the star is very far away. There are some Cepheid variables with luminosities of almost a million times that of the sun. Cepheids have been used to measure distances on the order of 1 million light years.

Let me show you a couple of examples of the first known Cepheid stars. They are super-giants that are near the end of their lifetimes.

Star	I (rel. to Vega)	Period (days)	Luminosity	Dist (LY)
Delta Cephei	0.024	5.366	2000	887
Zeta Geminorium	0.027	10.148	2900	1183

Table 11.2

Two Cepheid variable stars which are both more than 2000 times as luminous as the sun. The pulsation period of the most luminous Cepheid variables can be as long as a few years.

The luminosity of a common wax candle is almost the same for every candle. In fact a unit for luminosity is *candle* power. Stars like Delta Cephei are called *standard candles* because we can determine their luminosity from their period of fluctuation. If we can find standard candles (Cepheid variables) in distant galaxies, we can determine their distance by measuring the light intensity that we receive.

The largest standard candle

Mr. T.   In order to measure really distant objects, we need something that is incredibly luminous.

K.   We know that supernova are very bright. There must be big differences depending on how big the super--giant was before it blew up.

Mr. T.   There is, fortunately, another type of supernova.   It is a Type Ia supernova.   The designation for an exploding super-giant is a Type Ia supernova. Type Ia explosions are all remarkably the same.

J.   Different stars blow up and I don't see why they should all be the same size. How can these explosions have the same luminosity?

Mr. T.   You put your finger on the important feature of the pre-nova star- its size. They are all the same size. We have to examine more closely what is happening to understand why.

First, all Type Ia supernovae begin as white dwarfs.

K. I don't understand. You told us that energy production in white dwarfs had stopped and the temperature was too low for fusion to occur. How can a white dwarf explode?

Mr. T. It is a very unusual situation. In the formation of stars it happens frequently that two stars are formed together and do not combine. They rotate around each other. Let's look at the example of Sirius which is one of the brightest stars in the sky.

	T (deg. K)	L	Mass	Size	D (LY)	Separation (au)
Sirius A	9,500	25.4	1.46	1.7	8.6	8.1 to 31.5
Sirius B	5,800	$\sim 2.5 \times 10^{-3}$	0.98	0.008	8.6	

Sirius is really two stars, one of which is a white dwarf. It started out as two main sequence stars and Sirius B had the larger mass and became a red giant and finally a white dwarf.

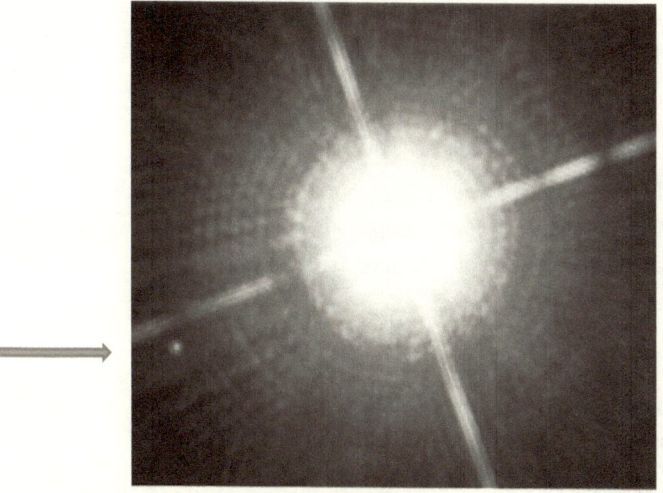

Fig. 11.7

Hubble picture of Sirius A & B. Sirius B is seen on the lower left of the picture.

K. Sirius B has less mass than A. How do know it started out with more mass?

Mr. T. Remember, large stars evolve faster than smaller ones. Sirius B lost mass by radiation and the formation of the planetary nebula that was ejected before it became a dwarf.

J.   I don't remember what an *au* is. Why does the separation have such a large range?

Mr. T.   An *au* is an astronomical unit. It is the distance from the Earth to the Sun. The large range from 8.1 to 31.5 *au* is because the orbit is elliptical and not circular.

Jill look at the Hubble picture in figure 11.7 and tell me what you think the next step will be in this star's evolution.

J.   I don't think Sirius B will do anything now, but Sirius A will grow like our Sun did in the simulation. It will become a red giant.

Mr. T.   Correct! Kay, look at the picture and tell me how big you think it will get.

K.   It is bigger than our Sun so when Sirius B's orbit is closest it might hit the outer reaches of Sirius A.

Mr. T.   Jill, what happens then?

J.   The dwarf has about the same mass as our sun but it is more than 100 times as small. The gravitational forces at its surface will be enormous. It will suck up material from the red giant. Its mass will increase.

Mr. T.   What will happen to its temperature?

K.   As the material is pulled in it will lose gravitational potential energy which will increase the temperature of the white dwarf.

Mr. T.   It will eventually reach the critical mass limit worked out theoretically by an Indian physicist by the name of **Chandrasekhar**. It is only 1.4 times the mass of the Sun. When that happens, the temperature will be high enough to cause Carbon and Oxygen fusion. This will put so much energy into such a small body that it goes boom. The flash of this explosion lasts only a matter of days. The gamma ray blast may only last for a few minutes. You really have to be on your toes or you will miss it. Kepler was able to see a supernova in 1604 and we can see its remnant today. The spectroscopic study of its remnant determined that it was a Type Ia supernova.

K. How does the spectrum show us the type of supernova?

Mr. T.   White dwarfs don't have a lot of hydrogen but they have Silicon which shows up in the spectrum. Type Ia supernova which come from collapsing giant stars have different spectroscopic signatures.

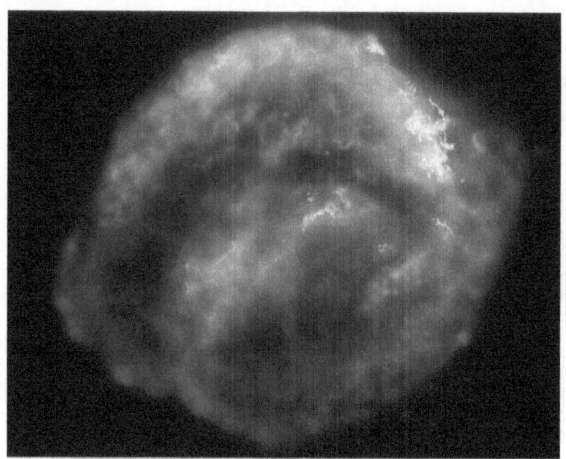

Fig. 11.8
Remnant of SN1604
This was the supernova observed by Kepler in 1604

J. Back to the main point of all this, tell us again why Type Ia supernovae are the biggest standard candles for measuring distance?

Mr. T.   I think you can see that if white dwarfs increase in size until they reach the Chandrasekhar limit, they will all be about the same size when they explode. Each explosion will be similar to the others and the light emitted will be enormous. Its luminosity can become 100 million times as great as our Sun's. This bright light allows us to see distances on the order of billions of light years. Light from this far away was produced near the time of the beginning of the universe.

That should be enough for today.

Session 12
You have stars in your eyes

"The nitrogen in our DNA, the calcium in our teeth, the iron in our blood, the carbon in our apple pies were made in the interiors of collapsing stars. We are made of starstuff." — Carl Sagan, Cosmos

In the classroom
More on supernovas

Mr. T.   Last time we spent some time on supernovas.   However the main emphasis of the session was on measuring distance. We have not exhausted your ability to understand these fantastic events and there are some important and exciting aspects that you should try to comprehend. We will spend more time examining what happens at the end of the supernova.

J.   Last time you told us about two types, type II and Ia. Are we going to learn about other types?

Mr. T.   I think it would be a good idea if, today, we concentrate only on type II. Unfortunately, we can't do everything. Let's start with stars more than 20 times the mass of our Sun. These stars have a surface temperature near 30 thousand degrees K. Kay, what do you remember about stars of this size from the last session?

K.   They don't live very long and after leaving the main sequence they move horizontally across the Hertzsprung-Russell diagram. This means they don't change their luminosity very much even though their surface temperature changes.

Mr. T.   Jill, what else did we discuss?

J.   They first became blue, yellow, and finally red super giants before blowing up.

Mr. T.   The blue supergiant is formed when the star burns nearly all the Hydrogen in its core. It contracts, heats up and Helium fusion is ignited.

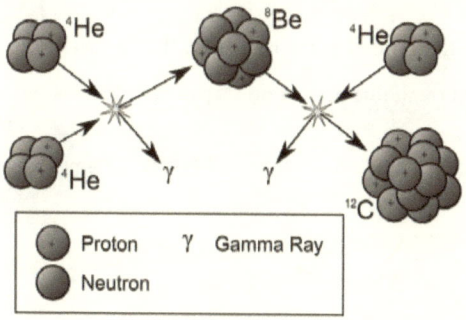

Fig. 12.1

Fusion of three alpha particles into a Carbon nucleus

J.   If its temperature increases, why doesn't it get even bluer?

Mr. T.   The outer shell still contains some Hydrogen and expands. This shell cools during expansion and the star changes color. Hydrogen is still burning outside the core.

When the Helium in the core is gone, the core contracts, heats up, and Carbon fusion is ignited. The outer shell expands more and both Hydrogen and Helium are burning outside the central core. The extreme outer shell becomes cooler. When the Carbon in the central core is used up, the process continues with Oxygen and then Silicon.

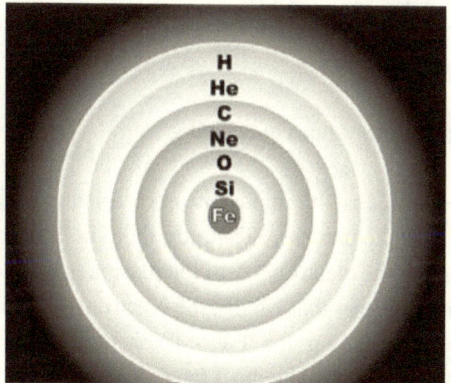

Fig. 12.2

Layers of a 25 solar mass star (not to scale)

J.   Does it just keep going through heavier and heavier elements?

Mr. T.   No, when the core is almost all Iron, the fusion in the core stops.

Core-burning nuclear fusion stages for a 25-solar mass star

Process	Main fuel	Main products	25 $M_\odot$ star		
			Temperature (Kelvin)	Density (g/cm^3)	Duration
hydrogen burning	hydrogen	helium	7×10^7	10	10^7 years
triple-alpha process	helium	carbon, oxygen	2×10^8	2000	10^6 years
carbon burning process	carbon	Ne, Na, Mg, Al	8×10^8	10^6	10^3 years
neon burning process	neon	O, Mg	1.6×10^9	10^7	3 years
oxygen burning process	oxygen	Si, S, Ar, Ca	1.8×10^9	10^7	0.3 years
silicon burning process	silicon	nickel (decays into iron)	2.5×10^9	10^8	5 days

Table 12.1
Nuclear fusion stages for a large star

K.   Why does it stop with Iron?

Mr. T.   Fusion gives us energy when two nuclei combine and the mass of the result is less than the sum of the combining nuclei. Einstein tells us that the lost mass is converted to energy. $E = mc^2$

When nuclei with mass greater than Iron combine, the total mass of the combination is larger than the parts that formed it. It takes energy to create the needed mass and put the nuclei together. Reactions that require energy don't happen automatically.

J.   When the core is all Iron does everything stop?

Mr. T.   I am afraid not! Gravity makes the core collapse and no fusion reaction is available to stop the process. The only thing holding back the collapse is the high pressure in the electron gas of the plasma that makes up the star. The density becomes so great that finally the electrons will not fit into the space available without violating the Pauli Exclusion Principle. Electrons can occupy the same space but they must have different wave properties in order to avoid interference. The description of this gas requires not only the use of quantum mechanics but relativity as well.

Eventually the pressure becomes so great that electrons are captured by the protons. This process changes the protons into neutrons and ejects neutrinos. The loss of the electrons reduces the pressure in the core and the collapse is accelerated. A huge blast of neutrinos occurs but only lasts for a few minutes. Nuclei are now falling toward the center at velocities of 23% of the velocity of light. The temperature in the core can now be as high as 100 billion deg. *K* and the luminosity much greater than all of the galaxy. The luminosity can exceed 5 billion suns but the brightest flash lasts only a matter of days. The core consists of neutrons which can't collapse further.

The sudden stop in the collapse causes the incoming material to bounce back from the center forming an outward moving shock wave. <u>This shock wave of neutrons crashes into the outer shell with enough energy and pressure to cause the formation of heavy nuclei (elements)</u>. Everything in the shell is blasted into space by this shock wave and forms star dust. In order to understand the dynamics completely, we need more information about the neutrinos that occur when the electrons are captured by the protons. Scientists believe that they play an important part in the formation of these shock waves. We were able to detect these illusive particles from SN 1987a. This supernova was not only very recent (1987) but fairly close (168,000 *Ly*). It could be seen with the naked eye. However, neutrinos are very difficult to study and the theory of a supernova shock wave needs more work.

J.   Did you say 168,000 light years was close? It doesn't seem close to me.

Mr. T.   We expect less than two supernovas per century in our galaxy. SN1987a occurred in our closest galactic neighbor the Large Megallanic Cloud.

Fig 12.3
The large Megellanic Cloud

Fig. 12.4
Hubble picture of remnant of SN 1987a twenty years after the explosion
The Hubble telescope was not finished in 1987 when this supernova occurred.
The shock wave is seen as a ring.

K.   Why is the formation of heavy elements by this shock wave so important to us?

Mr. T.   If we didn't have a supernova, we wouldn't have heavy elements. If we didn't have heavy elements, we wouldn't exist. Our bodies require these elements. When our solar system was formed out of the dust in space, much of this debris came from some previous supernovas. Most of the very heavy elements have decayed but the lighter ones have long enough lifetimes to survive a long wait in space. Some of the very heavy elements like Californium we can manufacture, but we do not find them naturally. Californium has a half-life of 898 years which is long enough to study in the lab but too short to be found in space.

K.   The shock wave of the supernova appears to be a ring. I would have expected a sphere; the shock wave must go in all directions.

Mr. T.   It is a hollow sphere. It appears as a ring because the telescope is getting light from more material at the outside edge of the hollow ball than the front face.

K.   Are new elements being created in the shock wave we see in figure 12.4?

Mr. T.   The new elements have probably all been formed before this picture was taken. The material in the shock wave is cooling and soon it may be cool enough to allow the formation of atoms. This means electrons will be captured by the nuclei, and chemical reactions will shortly begin to occur. This will mean atoms and molecules will appear and not just nuclei.

K.   How can we see what chemicals (molecules) are formed in the remnant of a supernova?

Mr. T.   Each chemical molecule has a particular vibrational frequency and most of the frequencies are in the radio wave portion of the electromagnetic spectrum. If a particular chemical is being excited, it will radiate at its characteristic frequency. If we know the frequency, we can tune a radio telescope to that frequency and look at an area of the sky where we expect these chemicals to exist. Specifically, we want to look where there is gas and dust. This is just like tuning to a radio station.

K.   What kind of chemicals have they found?

Mr. T.   The mix of chemicals has been described as "a pot of poison with a dash of sugar and a pinch of salt". Molecular hydrogen, water, carbon monoxide, ammonia, formaldehyde, acetone, hydrogen cyanide, sodium chloride, and ethyl alcohol are just a few of the chemical species identified. A whole new field of study called **astrochemistry** has been created. It is very exciting; many of these substances are required for life to form.

J.   Does this mean there could be life on other planets?

Mr. T.   We now speculate that there may be as many as 4 billion Goldilocks planets in our galaxy alone. We don't know and probably might never know if life exists on any of these planets. What do you think?

K.   Why did you call them Goldilocks' planets?

Mr. T.   When Goldilocks was eating the bears' porridge, she said this one is too cold; this one is too hot; but this one is just right. The one that is just right is a Goldilocks planet.

J.   Mr. Tweed, you told us in the last session that large stars become black holes. Will SN 1987a have a black hole at its center?

Mr. T.   The astronomers don't think so. Measurements indicate that it is probably a neutron star. However, if the models are correct, all stars that have a mass greater than 20 suns will become black holes. Astronomers have concluded that black holes exist in some locations based on the movement of visible objects (like stars) around an "invisible" gravitational source of considerable mass. To cite one example, the center of the Milky Way is the home of a supermassive black hole based on the observations of some stars nearby. This black hole is called Sagittarius A or Sgr A.

K.   What do you mean by supermassive black holes? How big are they?

Mr. T.   Do you mean, how massive are they? Specifying the physical size may be a problem. However, the radius of Sgr A's *event horizon* is only 0.08 *au*. This is called the **Schwarzchild** radius and theoretically it only depends on the mass of the star.

J.   Mr. Tweed, how do astronomers figure out all this stuff? You have spent quite a bit of time with distance but there is obviously a ton of stuff you have not explained. You have described some interesting phenomena but I don't have a good feeling about how we know these things just by looking through a telescope.

Mr. T.   I understand your problem and, unfortunately, I can't fix it. The only thing I can do in our short time together is pick out a few interesting techniques and try to give you some feeling about how these techniques can be applied to learn about the universe.

K.   You mentioned supermassive black holes and said that we have one in the center of the Milky Way. Is it possible for you to explain to us how we know there is a black hole and how we can determine its mass? I hope this is not too difficult for us to understand.

Mr. T.   I think we can do it. You first need to know a little bit more about stellar spectroscopy. You know already that stars emit light in the form of a continuous spectrum. The light from our Sun contains all the visible colors. However, outside the hot burning plasma there is a layer of cooler atomic gas. When photons with exactly the correct energy to excite electrons to a higher energy level move through this gas, they will be absorbed. We see dark lines in the continuous spectrum.

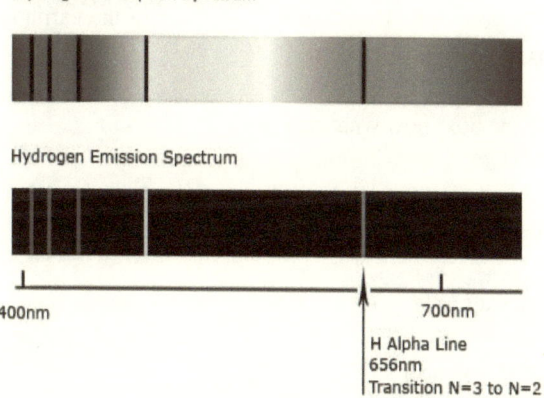

Fig. 12.5
A stars continuous spectrum showing hydrogen absorption lines.
Hydrogen emission spectrum shown for reference.

When these dark lines are examined, the composition of the gas is revealed. This in turn allows us to classify the star. The composition of the gas surrounding a blue supergiant will not be the same as a star of the same color in the main sequence.

Spectroscopic information may also help in determining mass. I will let Fritz help me at this point.

In the simulator

F. Hi! I am ready for you. Are you ready?

J. I don't know what to be ready for. What are we doing today?

F. Mr. Tweed showed you that there are some specific light frequencies that we can observe coming from a star. An example is the α line of Hydrogen. It has a wavelength of 6.56×10^{-7} *m* and a frequency of 4.57×10^{14} *Hz*. We can separate the light from a star, put it through a spectrometer, and observe a dark line at this wavelength. Today we are going to have a simulation that will help you to understand some of what we can do with these observations.

In our simulation, we are going to use sound waves instead of light waves. I have prepared a small box that emits a sound at exactly 1000 *Hz*. Consider this analogous to the α line of Hydrogen. Jill, take the box and put this earphone in your ear. I can talk to you through the earphone without Kay hearing us.

J. Ok, I have the box; now what do I do?

F. Switch on the box and take the little car and drive it at a steady speed down the road leading directly away from us.

J. I am on my way.

F. Kay, what do you hear?

K. Before Jill left I heard a steady high pitched sound. However, when she drove off, the pitch became lower and now it is getting fainter. I am unable to hear her at all now.

F. Kay, can you explain any of this?

K. The sound intensity is getting less because her distance is increasing. $I \propto \dfrac{1}{D^2}$ The pitch has gone down because of the Doppler Effect. We learned about that in session 6.

F. Kay, can you figure out her speed?

K. I would need to know more. I don't know the speed of sound, the reduced frequency of the sound coming from the box, and the Doppler equation.

F. I can help with all these. Take the large parabolic dish with the microphone at the focal point and point it down the road.

K. Where are the headphones? I need something to hear the sound.

F. You have two meters in front of you. The first reads the intensity of the sound and the second the frequency. Adjust the direction until the intensity is maximized. What does the frequency read?

K. It reads 963 *Hz*.

F. Jill, you have gone far enough. You can stop now.
    Kay, the speed of sound (*c*) is 333 *m/s*. Let's figure out the Doppler formula and you can then tell me Jill's speed. My black box emits the crest of a wave which will normally travel one wavelength (λ) toward us before the next crest is emitted a time *T* later. In this case Jill travels a small distance (*vT*) before the next crest which increases the wavelength to λ_2.

$$\lambda_2 = \lambda + vT$$

If this is solved for *v*, we get: $v = \lambda_2 f - \lambda f$,   because $f = 1/T$.
The wavelength and frequency are related by: $\lambda f = c$ and $\lambda_2 f_2 = c$ which can be used to simplify the velocity result.

$$v = c \times \left(\frac{f}{f_2} - 1\right)$$ (12.1)

K. I now know f, f_2, and *c*. Jill's velocity is, therefore, given by:

$$v = 333 \times \left(\frac{1000}{963} - 1\right) = 12.8 \frac{m}{s}$$

Which is about 46 *km/hr*.

F. It is probably worth looking at equation 12.1 for a minute. If the frequency received is equal to the one transmitted (1000 *Hz*), the quantity in parentheses is zero and the calculated velocity is zero. If the received frequency is greater than 1000 *Hz*, the quantity in parentheses becomes negative which means the car is coming toward us.

Kay, pardon me while I talk to Jill. I will only be a minute.

Jill, you will find a cord 3 meters long in the car. Attach it to the box and swing it in a horizontal circle as fast as you can. Try to keep the speed constant.

J.   How much longer is this going to take? I am getting sick of listening to this loud note.

F.   Not long, we are nearly finished.

Kay, what's happening now?

K.   The pitch is moving up and down. The highest frequency is 1028 *Hz* and the lowest is 974 *Hz*. The period of the fluctuation is just over 2 seconds.

F.   Is Jill moving away or coming closer?

K.   The pattern is repetitive so I don't think she is continually moving closer or further away. It looks like she moves closer and further in a repetitive fashion. However, the shift in frequency is similar to what we measured when she was moving away. 963 *Hz* is close to 974 *Hz* which means the maximum speed is fairly high. I don't think she could drive the car back and forth every 2 seconds and she can't run that fast. She must be doing something with the box that emits the sound. Fritz, I don't understand; help me.

F.   I will tell you she is not changing the frequency the box emits. What you are observing comes from some sort of motion. What sort of motion repeats itself on a regular basis?

K.   An oscillation or something moving in a circle.

F.   You got it! I will let Jill explain when she gets back.

Jill, you can come back now.

J. I'm back but what were you doing all that time. I was getting tired whirling that box around my head and listening to the constant sound.

K. I get it! When the box is moving toward us, the pitch will be higher and when it is moving away, it will be lower. Jill was whirling it around once every 2 seconds. Its speed should be constant but I only get a reading when it is moving toward or away from me. I will calculate its speed moving away.

$$v = 333 \times \left(\frac{1000}{974} - 1\right) = 8.9 \ \frac{m}{s}$$

F. Even though you couldn't see it, you have enough information to calculate the radius of the circle it was traveling in.

K. I don't understand. How can I do that?

J. I don't think it's too hard. The speed (v) is the circumference of the circle ($2\pi r$) divided by the period (T). $v = \dfrac{2\pi r}{T}$ or $r = \dfrac{vT}{2\pi}$

K. You are right, I will calculate the radius and you tell me if it correct.

$$r = \frac{8.9 \times 2}{2\pi} = 2.83 \ m$$

J. You are a little off. The correct answer is 3 m. However, you were only about 6% off and you didn't measure accurately the period of rotation. You said it was a little over 2 seconds but you used 2 in your calculation.

I am really amazed that we can measure a radius of an orbit using Doppler shifts. Can we do that with stars? Are there stars that orbit other stars?

F. There are many binary stars where they orbit each other, and also stars orbit black holes. There are a number of stars orbiting the large black hole in our Milky Way. If we can measure their orbital period and velocity we can calculate the radius of their orbit, and if we have this information, we can calculate the mass of the black hole. You only have to remember Newton's universal law of gravity, his force law, and the acceleration of an object moving in a circle. You learned about these in session 3. See if you can remember.

J.   The second one is easy. The force on an object is equal to its mass times the acceleration caused by the force. $F = ma$

K   The universal law of gravity states that every object in the universe attracts every other object with a force that is proportional to their masses and inversely proportional to the square of the distance between them.

$$F = G\frac{M_1 M_2}{R^2}$$

J.   I don't remember the formula for acceleration of an object moving in a circle.

F.   If you look back in session 3, you will find it, but I will save you some time and give it to you: $$a = \frac{v^2}{R}$$

K.   I think I get it! The force on object 2 is: $F = M_2 \times \frac{v^2}{R}$

If we equate this to the universal gravitation force, we get:

$$G\frac{M_1 M_2}{R^2} = M_2 \times \frac{v^2}{R}$$

M_2 cancels out and we can solve for the mass M_1.

$$M_1 = \frac{v^2 R}{G} \text{ or } \frac{4\pi^2 R^3}{GT^2} \tag{12.2}$$

We can determine v and R from Doppler shifts and the period (T) by monitoring how the Doppler frequency changes with time. We already know the universal gravitation constant G. I assume the period could be very long, even a number of years.

F.   We have some data that will allow us to calculate the mass of the black hole at the center of our Milky Way.

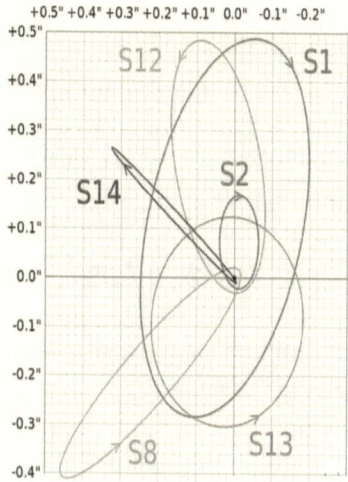

+0.5" +0.4" +0.3" +0.2" +0.1" 0.0" -0.1" -0.2"

Fig. 12.5

The orbits of several stars around Sgr A   S1 has a rotational period of 94.1 years and an orbital radius of 3300 au. The dimension of the x and y axes is given in parallax arc seconds.

The diagram shows the orbits of a number of stars around the black hole. We know both the period and the radius (semi-major axis of ellipse) of S1's orbit, therefore, we should be able to calculate the mass of the black hole from equation 12.2

J.   How do we know the period is 94.1 years?   We have only made measurements for a few years.

F.   If we measure its position and velocity (Doppler) for only a few years, we have enough information to calculate its orbit.

K.   If we want to use equation 12.2 $\left(M_1 = \dfrac{4\pi^2 R^3}{GT^2}\right)$, we will have to convert the data into MKS units.

$$T = 94.1 \; years \; \times \frac{3.16 \times 10^7 \, seconds}{1 \; year} = 2.97 \times 10^9 \; seconds$$

$$R = 3300 \; AU \times \frac{1.50 \times 10^{11} \, m}{1 \; AU} = 4.95 \times 10^{14} \; m$$

If I put these numbers into the equation I get:

$$M = \frac{4\,\pi^2 \times (4.95 \times 10^{14})^3}{6.67 \times 10^{-11} \times (2.97 \times 10^9)^2} = 8.3 \times 10^{36}\ \textit{kg}$$

J.   That looks like a big number but I don't know what it means. How big is it?

F.   The mass of our sun is 1.989×10^{30} *kg*. This means the black hole Sgr A is about 4 million times as massive as our sun.

Gravity inside a large mass

F.   If you are happy with these concepts it is time for a much different simulation. Please step into the elevator behind me.

J.   I am going to assume that this is not your everyday type of elevator.

F.   You are correct. This elevator will take you to the center of the earth.

J.   Will it be able to withstand the high temperature and pressure?

F.   Don't be silly; this is a simulator. Down we go. What do you feel?

K.   I feel a little lighter but I assume it's because the elevator is accelerating downward.

F.   Good! The acceleration will stop in a few seconds and we will be traveling at a constant speed. We are now traveling very fast and approaching the center. How does it feel now?

J.   I feel much lighter. Please explain why?

F.   Gravity from the whole earth is acting on you but some is pulling up and some is pulling down. The net result is less pull; you feel less weight. We will now decelerate and stop at the center.

J.   Fritz, you should warn us; we were almost floating and the deceleration made us fall to the floor. I feel really heavy. We were lucky no one was hurt.

K.   I think the elevator has stopped. This is really weird; we are just floating around. There is no gravity at all.

F.   When you are at the center gravity pulls in all directions equally. All these forces add up to zero. I won't go through the math but I can help you a little to understand what is going on. When you are a distance R from the center all the mass closer to the center than you contributes to the gravity you feel. The gravity from all the mass more distant than R from the center cancels out.

J.   Do we have to go back now? It's fun to just float around.

F.   Yes we must go back and you need to remember what we did here in the next simulation. We will now travel to the galaxy Andromeda.

Dark Matter

Fig. 12.5
The galaxy Andromeda
Dist-2.5 million LY, diam.-150,000 LY, 10^{12} stars

J.   Is this going to be one of those long trips like our California excursion?

F.   Not at all. We will have a comfy space ship with lots of instruments and controls and we can see in all directions. We will be able to adjust our clocks so that we can observe events that take a long time and be able to measure distances with only a mouse click. We will confine our investigation to within the galactic disk. Are you ready?

J.   You will warn us if we are going to be tossed around like we were in the elevator.

F.   Don't worry, we will now travel to the central bulge about 5 *LY* from the center.

K.   It is very bright here; there are so many stars. The sky is just full of them.

F.   See the very bright one in front of us. It is orbiting the galactic center and we are in the same orbit. At the present time this star is moving toward Earth. Unfortunately, it can never move exactly toward earth because Andromeda is tilted slightly. Jill, read the velocity meter and record how fast we are moving.

J.   It reads 5.625×10^4 *m/s*. Is that very fast?

F.   It is fast enough to produce a small but significant Doppler shift in its spectrum toward the blue end. We can compensate for Andromeda's tilt with instruments on earth and calculate the star's velocity. Let's move out to 20 *LY*. Read the velocity meter again.

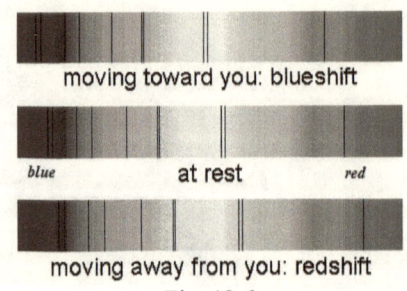

moving toward you: blueshift

blue   at rest   red

moving away from you: redshift
Fig. 12.6
Doppler shifts in the dark spectra from a star

J.   It reads 2.25×10^5 *m/s*. That is much faster. We just learned that the mass of an object can be determined from the velocity and radius of an orbital object. The result was given by equation 12.2 ($M_1 = \dfrac{v^2 R}{G}$). If both *v* and *R* are increasing, the mass of the galaxy must be increasing.

F.   You need to remember the elevator simulation. What happened to gravity when you went deep into the earth?

J. I remember now; when we went into the earth the only part of the earth's mass that affected us was the part closer to the center than we were. When we are in the galaxy the mass of all the stars further out from the center than us will cancel. As we move out, more and more stars will act gravitationally on us. If we went to the center of the galaxy, there would be no gravity.

F. If you went to the center of Andromeda, you would have a big problem because you would find a black hole with 100 million (10^8) times the mass of our sun. Otherwise your analysis is correct; good job.

Let's continue moving out, following stars, and recording the velocity.

K. This is kind of fun but we have a problem; I can see the galaxy on our left but I don't see any close stars on our right. We have moved out to 75,000 LY and there are no orbiting stars to follow. I looked up Andromeda and read that it is only 150 LY in diameter. What do we do now? Is it time to quit and go home?

F. Not yet! We have one more trick up our sleeves. In the outer reaches of the galaxy there is a quantity of orbiting dust. On earth this dust can be detected with radio telescopes and Doppler measurements can be made on the radio signals from this dust. I will adjust the sensors on our spaceship so that we can follow the orbiting dust out to about 100 LY.

K. There is something wrong. The velocity is not changing. According to equation 12.2 ($M_1 = \dfrac{v^2 R}{G}$), if the velocity is constant and the orbital distance (R) is increasing, the mass of the galaxy must be increasing. However, there is nothing there that we can detect. The velocity should be going down as we move away from the center. Fritz, please explain.

F. Let's look at the data we have collected in the form of a curve and see what we can figure out.

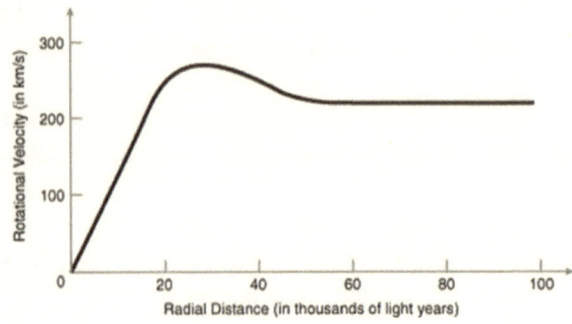

Fig. 12.7

Circular orbital velocity as a function distance from the center of Andromeda

J.   Kay is correct; the velocity is not decreasing as we move out; it seems to be almost constant. Maybe it will drop if we move out even more.

F.   Unfortunately, that is not possible; we have run out of stars to follow and there is not even any observable gas or dust. We have no way to make any measurements from earth using the Doppler shift techniques we have discussed.

K.   So what conclusions can we make from all of this?

F.   The mass of this galaxy that we can't see we call **Dark Matter**. It doesn't react with electromagnetic energy in a normal way, which is another way of saying we can't see it. It's dark! We don't know what it is and understanding it has become one of the most elusive topics in all of science. It is a great mystery!

However it is showing up in lots of sophisticated astrophysical observations. Einstein's general theory of relativity shows us how gravity can bend and distort space, and mass produces gravity. We are seeing more and more evidence of space distortions that are attributable to dark matter.

Astronomers now calculate that only 4.6% of all the mass in the universe can be observed optically. 23.3% is dark matter and 72.1% is dark energy.

Dark energy will be one of the topics in the last session, next time. See you then.

Session 13
The Universe and the Great Unknown

"The known is finite, the unknown is infinite; intellectually we stand on an islet in the midst of an illimitable ocean of inexplicability. Our business in every generation is to reclaim a little more land." — T.H. Huxley

In the classroom

Mr. T.   Today is our last session and we will take a quick look at the universe and some of the unsolved mysteries of science. In 1900 we didn't know very much about the universe. We had very little idea how far away the stars were, what they were made of, how they burned, how hot they were, if they moved or were stationary, or how old they were. We also had no knowledge about galaxies, nebulae, quasars, pulsars, and a host of other phenomenon.

Our present knowledge of the universe would come in a deluge of theoretical and observational progress that started at the beginning of the twentieth century. There would be many great scientists involved in this quest for knowledge and understanding, but the star of this effort would be Einstein.

Einstein's Universe

K.   You don't really expect us to comprehend Einstein's Universe, do you? I understand that it's all about general relativity. You gave us a short introduction to special relativity and hinted at a few things that general relativity explained, but you didn't say anything about the fundamental difference between special and general relativity.

Mr. T.   First, I don't expect you to gain a good comprehension of the general theory. The subject is difficult for almost everyone. Even Einstein had serious problems with it. Einstein called his special theory of relativity (finished in 1905) "*child's play compared with what came after*".

The special theory involves only objects moving at constant speed. The general theory considers also acceleration and gravity.

J.   I can see why he should include acceleration which is needed to discuss forces, but why should gravity be included?

Mr. T.   I can't explain very much but I can give you a few hints as to what is involved. Let's look at a version of one of Einstein's thought experiments.

You wake up in a windowless space craft, sitting comfortably in your chair and feel your weight pushing you into the chair. What can you conclude about your situation?

J.   You are probably on the ground and have not taken off yet. If you were in space, you would be weightless.

K.   That's possible, but you could also be in zero gravity and accelerated by the rocket engine. You would feel the force of the chair causing you to accelerate. Your acceleration would be close to 1 *g* if you feel your normal weight.

Mr. T.   Can you tell which explanation is true? Who is correct: Kay or Jill?

K.   I don't think it is possible to tell without windows or some other measuring technique.

Mr. T.   Einstein would agree with you. He claimed the two situations were equivalent and proceeded to work out a theory based on this ***principle of equivalence***. We will not go into the details of his general theory but I will list some of the results.

Your clock speed compared with my clock will depend on the speed of your motion with respect to me.
Your clock speed compared with my clock will also depend on your acceleration.

Your clock speed compared with my clock will depend on differences in gravity.

The distances you measure compared with my measurements will depend on differences in gravity, motion, and acceleration.

Therefore, space-time is curved and Euclidian geometry will not work when either gravity, acceleration, or motion is extremely high.

J.   It's just a theory; how do we know if it is true or not?

Mr. T.   *A scientific theory is a well-substantiated explanation of some aspect of the natural world, based on a body of facts that have been repeatedly confirmed through observation and experiment.*   When a theory is sufficiently confirmed it becomes *a scientific theory* which is much more than a hypothesis. We can ask: is Newton's theory of gravity true or not? It will not explain as many situations as Einstein's theory so maybe it's not quite as true as the General Theory of Relativity. However, Einstein's theory does not explain some things, such as what goes on inside a black hole. We need to understand the limitations of these theories but asking if they are true or not is a very philosophical question. I don't think it would be correct to say that either one is false.

After formulating his general theory, Einstein waited a few years and then published a paper on cosmology. This was not easy work for him. Max Plank warned him against working on general relativity: "*As an older friend I must advise you against it for in the first place you will not succeed, and even if you succeed, no one will believe you*". He persevered but begged his friend Marcel Grossman (mathematician): "*You must help me or else I'll go crazy!*" In another letter he worried that he had "*again perpetuated something about gravitation theory which somewhat exposes me to the danger of being confined to the madhouse*".

J.   Is Einstein's General Theory of Relativity now considered a scientific theory or is it still just a theory?

Mr. T.   It is definitely a scientific theory and is used every day in a great many practical ways. To name a couple: If the satellite clocks used for GPS navigation were not corrected using general relativity the system would not work properly. Most of the analysis of the experiments done in Cern Switzerland with particle accelerators also require the theory.

After applying his theory to the cosmos, he concluded:

> The universe is finite.
> The universe looks generally the same for all observers. This idea is
> called the ***cosmological principle.***

J.   I thought our universe was infinite in size and the stars just went on and on forever.

Mr. T.   We can make a good argument that an eternal universe of infinite extent does not exist. Let's assume for a moment that the universe was homogenous, eternal and infinite in extent. If that were true, we would be blinded by light. There would be no day and night and the sky would be bright white.

J.   How can you say that? Most of the stars and galaxies would be far away and very dim.

Mr. T.   That's true, but as you consider greater distances, the number of stars would increase as the cube (volume) of the distance ($I \propto d^3$) and the light dims only as the square of the distance ($I \propto \frac{1}{d^2}$). As you consider more stars and greater distances, the intensity of light would increase ($I \propto \frac{d^3}{d^2} \propto d$). As the light is calculated from more stars at greater and greater distances, the calculated value would rise without limit. Conclusion: we don't have an infinite, eternal, and homogenous universe.

K.   I don't understand how the universe can look the same for all observers; if you lived at the edge of the universe; you would see all the stars and galaxies on one side and just space on the other. Is that not true?

Mr. T.   According to Einstein there is no edge to the universe. The universe is composed in such a way that every person can consider himself at the center. No matter where you are, you can look out and see galaxies in all directions. His view of the universe is definitely not intuitive.

J.   I know that it is not within our means to understand all of Einstein's ideas but you are not outlining in a very definitive way how he was able to come to these conclusions. Can you help us gain a better feeling for this material?

Mr. T.   Many people have written complete books on this subject in order to help non-scientists gain some sort of feeling for this material. Unfortunately, even though these efforts help to some extent, most readers are left almost totally confused. The one positive thing that happens is they gain more faith in the scientists and become more accepting of the theory.

J.   Are you expecting us to have blind faith in the scientists without fully comprehending their ideas?

Mr. T.   Unfortunately, modern science has become complicated enough that most people are not able to understand more than a small amount. The details are overwhelming to most of us. About the only thing we can do is listen to the predictions of these scientific theories and pay attention to the experimental and observational verifications of these predictions. In many cases, you will be able to say: the theory seems to be verified and it help us understand our world. In other cases you might say: I am not sure they have made such a good case; I think I will wait for more corroboration.

When Einstein first worked out his picture of the universe, he was not happy. I think we can figure out what bothered him. His picture of gravity is different than Newton's but both show masses attracting, even at large distances. What do you suppose Einstein would predict about what all the masses in the finite universe would do to each other?

K.   They would attract each other.

Mr. T.   What would this attraction do over time?

K.   They would accelerate toward each other and eventually come together.

Mr. T.   Exactly! This is called the **big crunch** and he didn't like it. He even invented a trick to avoid it; he added a constant to his equations called the **cosmological constant** so that the equations would not predict it. He later called it the greatest blunder of his entire life.

J.   Why did he consider it such a mistake?

Mr. T.   I don't exactly know, but perhaps it was because he imposed his own preconceived view on the science. He believed the universe was eternal and static.

Hubble

In the early 1920's a new scientist made his contribution to our understanding of the universe. **Edwin Hubble** found a Cepheid variable in the Andromeda Nebula and was able to measure the distance (900,000 *LY*). He identified Andromeda as a galaxy and not just a nebula. This instantly expanded our understanding of the size and extent of the universe.

He next made his greatest contribution by finding numerous other galaxies and observing a red shift in each of their spectra.

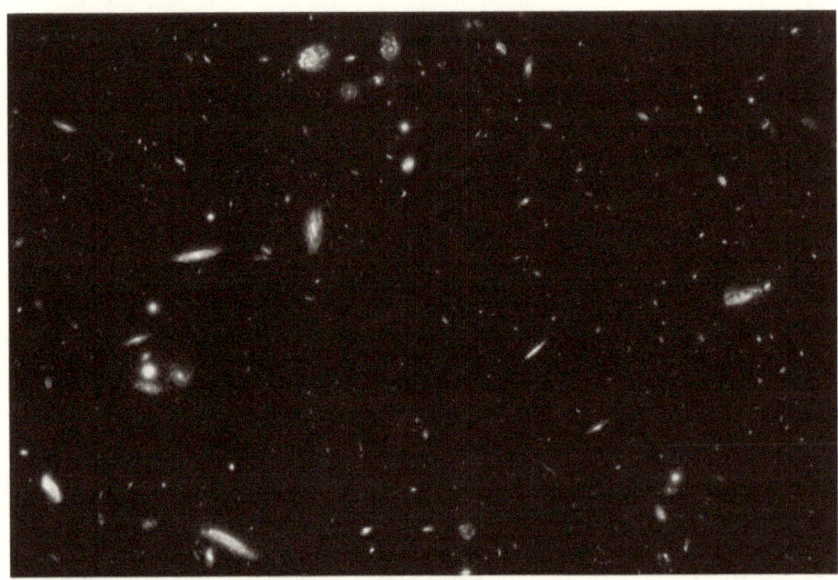

Fig. 13.1

NASA's Hubble Space Telescope reached back to nearly the beginning of time to sample thousands of infant galaxies. This image, taken with Hubble's Advanced Camera for Surveys, shows several thousand galaxies, many of which appear to be interacting or in the process of forming. Some of these galaxies existed when the cosmos was less than about 2 billion years old.

K.   The red shift means they are moving away from us, doesn't it?

Mr. T.   Yes! They were not only moving away from us but their speed (*V*) appeared to be proportional to their distance (*D*).

$$V = HD \tag{13.1}$$

The constant of proportionality *H* is known as the Hubble constant.
Our universe was expanding. Einstein would say space was expanding.

Lemaitre

A priest and member of the *Pontifical Academy of Sciences,* **Monsignor Georges Lemaitre** reasoned that our expanding universe would have galaxies racing away at speeds proportional to their distances from us. He and a Russian mathematician, **Alexander Friedmann,** actually separately calculated this same result using the General Theory two years before Hubble made his measurements. His idea was that if the universe is expanding it must have been very compact at some point it time. He called it "*a primeval atom of small but finite size*". Einstein had earlier rebuffed Friedmann and now did the same to Lemaitre. This was the second time Einstein rejected the idea of what was later to be called *The Big Bang*. At this point most scientists continued to believe in an eternal static universe.

K.   How could Lemaitre calculate this from the general theory?   I thought Einstein introduced the *cosmological constant* to stabilize the universe.

Mr. T.   Lemaitre ditched the *cosmological constant* and proceeded without it. After Hubble published his results showing the universe was expanding, Einstein also abandoned it.

Gamow

George Gamow considered Lemaitre's big bang but wanted to start with a dense soup of photons, protons, neutrons, and electrons. His goal was to show how the elements could be synthesized from the intense pressure and heat at the beginning of the big bang. Unfortunately, his mathematical skills were not up to the difficult task of modeling this hot dense soup. He was, however, able to find a very capable mathematician graduate student named Ralph Alpher. Together, they worked for several years on the problem of nucleosynthesis. They were able to predict with remarkable accuracy the amount of hydrogen and helium in the universe. Gamow wanted to show how all the elements were formed in the big bang but was unable to do so. We now know that the stars and novae are the factories for element fusion and not the big bang.

K.   Mr. Tweed would you explain why the intense heat and pressure of the big bang was not able to cause fusion into larger nuclei. It seems to me that everything that was required existed in this soup; it was hot, under high pressure, and there were lots of protons to fuse.

Mr. T.   That's true, but what was needed was a Goldilocks soup: not too hot, and not too cold. Everything had to be just right. If the temperature was too high, the nucleons would bounce off each other. If the temperature was too low, they would not have enough energy to overcome the electrical repulsion and get close ($\sim 10^{-15} m$) enough to fuse.

Even if you had a Goldilocks soup, it would not stay that way for very long because the universe was in the middle of an explosion. It's estimated that the period of nucleosynthesis lasted only 17 minutes, and that was just enough time for the helium of our universe to form with only a tiny amount of heavier elements.

The contributions of Gamow, however, were not finished. He reasoned that as the soup cooled it would eventually reach the point where electrons and nuclei would combine and form atoms. With the unattached electrons gone, the universe would become transparent and the photons could travel relatively freely.

J.   I don't understand why attaching the electrons to the nuclei makes the universe transparent.

K.   I can answer that question. When the electrons are free a plasma exists just like in the center of our sun. Light can't travel easily through our sun. It interacts with the charged particles and gets bounced around.

J.   What has to happen for the electrons to be able to attach themselves to the nuclei?

Mr. T.   It turns out to be fairly simple. The temperature must drop to about $4000°K$. We can see this happen in the Corona of our Sun.

J.   How long did that take for the temperature to drop that far?

Mr. T.   Present models indicate that it took about 380,000 years from the initiation of the big bang. Gamow's contribution came when he realized that this radiation would have a continuous black body spectrum much like the

Sun's radiation and the radiation should still be around today; it should be everywhere in the sky.

J.   That's crazy, I look into the sky at night and all I see is the stars, I don't see any black body radiation.

Mr. T.   The universe has expanded a lot since then, and as it has expanded, the wavelength of the original radiation has gotten longer. The universe has cooled and the spectrum of the $4000°K$ radiation has shifted with it. The present temperature of this radiation is about $2.7°K$, which is very close to Gamow's calculation. This radiation is no longer visible or even infrared; it has been shifted into the microwave region and was not detected until recently.

Fig. 13.2
A complete map of the heavens (2012) showing the cosmic microwave background radiation. The pattern is almost perfectly uniform and the slight deviations have been enormously amplified in the map shown here.

It is beyond the scope of this class, but careful studies of this radiation pattern provide more than just confirmation of the big bang theory. They also give us considerable insight into the dynamics of the early universe.

J.   Why didn't the radiation just escape?

Mr. T.   Where is it going to go? There is nothing, by definition, outside of our universe. One of Einstein's predictions is that radiation is contained in the universe.

K.   Mr. Tweed, you told us that Einstein didn't like the idea of a Big Crunch or collapsing universe so he invented a cosmological constant to avoid it. Now that he has abandoned this constant, will we have a Big Crunch? Is the universe going to collapse?

Mr. T.   Around 1990 measurements began to indicate that the expansion of the universe was not slowing down as expected. It was actually accelerating! This would seem to indicate that our universe will expand at an ever increasing rate forever. Eventually it will freeze, disperse, and go dead.

K.   Why is it accelerating? What would cause that to happen?

Mr. T.   We don't know!   However, Einstein's cosmological constant has returned. If we give it a different value and return it to his equations, it describes the acceleration. It seems to describe a situation where space itself is filled with dark energy, and as space itself expands so does this dark energy.   This energy is causing everything to fly apart faster and faster.

J.   Mr. Tweed, now you getting really weird. I could almost follow the dark matter argument but dark energy is too much.

Mr. T.   I don't understand it. The scientists don't understand it. I would be extremely surprised if you said you understood it. In fact I would not believe you. We can make measurements; we have equations to describe it; we have a few weird ideas about it, but we don't understand it.

J.   If it is so weird, why do scientists believe it?

Mr. T.   The big bang theory predicted the cosmic microwave background. After some time, we found it and it fit the theory. The theory predicted the relative abundance of elements which has been confirmed with current spectroscopic measurements. The theory predicted an age of the universe. Distance and velocity measurements for galaxies allow us to work backward to a time when they were all together. This gives us an age for the universe which is nearly the same as the microwave background calculations. The age of our universe is approximately 13.7 billion years. The more predictions that are verified, the more credence we give to the theory.

J.   What came before the big bang? Do other universes exist?

Mr. T.   We don't know and we can't know. Some people have ideas about things like this but they will probably never rise to the level of scientific theories because there is no way to make any confirming observations or do any experiments. Having said that, it is still interesting to think about, isn't it?

There are many interesting things that we can't understand. It is probably worthwhile examining a few. Some of them may have been worked out by the time you read this but we may never figure out some of the others.

Matter-antimatter symmetry violation

Every particle has an antiparticle and when we observe the creation of particles using the accelerators at Cern or other particle research facilities, they are always created in pairs. In the example shown below a positron-electron pair is produced from a high energy gamma ray.

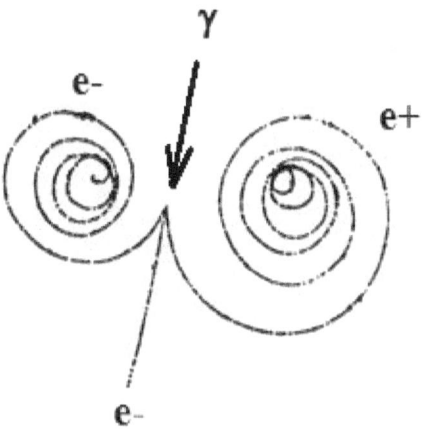

Fig. 13.3

Here an invisible high-energy electrically neutral light particle (γ photon) traveling down from the top of the image scatters off an atom (located near the middle of the figure leading to the creation of two new particles, a negatively charged electron and a positively charged antiparticle called a positron. The created positron's charge has exactly the same magnitude e as the created electron's charge but is opposite in sign so that the total charge has not changed The second electron (e-) was kicked out of the atom when the γ ray struck and was moving too fast for the magnetic field to bend its path substantially.

Our theoretical understanding of elementary particles predicts that during the big bang equal numbers of particles and antiparticles would be formed.

J.   So what is the problem? Why can't we have a universe with both particles and antiparticles? We are able to create them in the laboratory.

Mr. T.   There is a problem; when an antiparticle comes in contact with a particle they annihilate each other and high energy photons are produced.

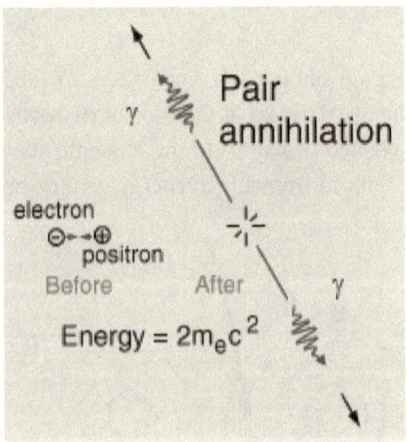

Fig. 13.4

Figure showing the obliteration of an electron-positron pair and the formation of two γ rays

K.   If all the particles destroyed all the antiparticles, we would have no matter at all.

Mr. T.   That's correct. We either have a problem with our understanding of matter creation or there must be a large amount of antimatter somewhere in the universe. I think it is time for your last trip to the simulator.

In the simulator

F.   Welcome back, you have been gone for a while. Today I have arranged for you to have a telephone conversation with someone very far away.

J.   Where is this person?

F.   He lives on a planet much like ours in the Andromeda galaxy. I think it would be a good idea to find out if he lives in an antimatter galaxy or one like ours. In 4 billion years or so our galaxy will collide with his. If Andromeda is an antimatter galaxy, the collision could be pretty spectacular.

K.   That's crazy! There is no way we can talk to someone 2.5 million *LY* away.

F.   You don't understand, this is a simulator and we can simulate anything. It doesn't matter whether it is ridiculous or not. The computer will translate the languages back and forth.

K.   What do we want to talk to him about?

F.   If you ask the right questions, maybe you can determine if his galaxy is made of antimatter. The laws of physics might be slightly different.

K.   Why can't we just examine the light coming from the stars in the galaxy? Maybe the spectra will be different in some way if the galaxy is composed of antimatter.

F.   That's a good idea but the energy levels in an antimatter atom should be exactly the same as with normal matter. Therefore, the spectra should be the same.

J.   OK, here goes, I am calling. Hello, my name is Jill. I hope you are having a good day. What is your name?

N.   My name is Neruf. Unfortunately, we don't have days on this planet. The same side of the planet always faces the star that gives us light and heat. Our sun is always shining where I live. The other side of the planet is dark and cold. No one lives there. However, I have been having a good time talking to Fritz. He has been explaining your laws of physics to me.

K. Hi Neruf, my name is Kay. I assume our laws are the same as yours. We need to design an experiment that will determine if you have positrons or electrons in your galaxy. If you generate a magnetic field pointing up and send horizontally one of your electrons, we can tell if it positive or not. If it is positive, it will move in a clockwise direction when viewed from above. Is it possible for you to do that?

N. After talking to Fritz, our laws seem to be identical. However, there are some problems. We don't have a common frame of reference. Your meter is roughly defined as one millionth of the distance from your equator to the North Pole. Our planet is not the same size. We can make an atomic clock but even that will read different time intervals than yours because of changes in gravity. It will be very difficult to establish the same units that you have. One of the biggest problems is: we can't determine if our right is the same as your right, and we also have trouble with clockwise and counterclockwise.

We have come to the conclusion that there is no way for us to determine clockwise from counterclockwise. We can't do your experiment. After discussing the problem with our scientists and with Fritz, we have decided that it is impossible for us to determine if our galaxy is made of antimatter or not.

However, we strongly believe that we both are made from matter and not antimatter.

K. If you can't do any experiment to answer the question, why do you believe so strongly in matter?

N. Fritz has explained how much you can see with your big telescopes. He explained that you have seen distant galaxies actually collide. If any matter-antimatter collisions ever happened, there would be so much radiation that it would make a supernova look like a tiny spark. You have never discovered anything like that, so we doubt that there is any appreciable amount of antimatter in our universe.

K. It was nice talking to you even though you could not answer our question. Good bye.

J. Fritz, we are not happy with you. You knew we would not get any positive answers from the telephone call and you did it anyway.

F. Don't blame me. I am innocent. I only do what Mr. Tweed tells me. In any case it should be clear to you that we don't understand why there is more matter than antimatter. It is part of the great unknown.

K. We won't see you again will we?

F. I don't know. Maybe someday. In any case learn a lot and have fun.

In the classroom

Mr. T. Are you ready for our next excursion into the great unknown?

J. Don't make it too complicated. My head hurts.

Mr. T. We have already discussed this topic, but I want to say just a few things before we leave it.

Dark Matter and Dark Energy

The idea of dark matter was introduced to explain why the material in the outer portion of galaxies rotates with a higher velocity than we expect. Other explanations have been put forth. Maybe Newton's law of gravity needs to be modified for the very low centripetal accelerations experienced as we move away from the galactic center. Most physicists don't like this approach. The equations get messy and scientists are always looking for beautiful simple solutions and not ugly complicated ones. There is even an unwritten rule: if there are two explanations for the same phenomenon and you can't decide between them, you must pick the simplest one. In any case dark matter remains within the confines of the great unknown.

I believe dark energy is even more of an enigma than dark matter. At first it was just a trick by Einstein to insure that the universe existed in a steady state. His cosmological constant has now been tweaked to explain the acceleration of the expansion of the universe. This process seems to predict an energy that grows as space expands and pushes things apart. Einstein tells us that anytime you have energy you have mass and anytime you have mass you have energy. $E=mc^2$ or $m=\frac{E}{c^2}$. If this idea is extended to our universe, we find that only a few percent of the total composition of the universe is normal everyday stuff.

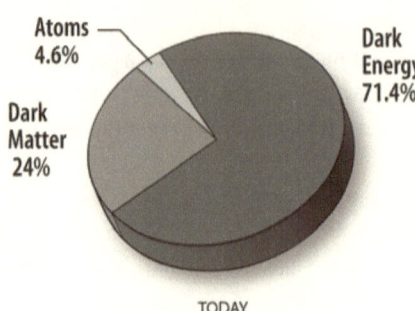

Fig. 13.5
Composition if our universe

J.   There seems to be a huge amount of dark energy. Why did it take us so long to discover it?

Mr. T.   We needed powerful telescopes and a rather sophisticated understanding of our universe before we could generate any ideas about dark energy. However, although the amount is large, the density is not. The amount of atomic mass and dark matter is concentrated and appears to be constant. Dark energy seems to take up all of space and as space expands so does the energy. In the future, the composition will be different than pictured above.

In any case, our understanding of dark energy is illusive and remains part of the great unknown.

K.   Do you think our ideas about dark energy are very solid? Are these ideas liable to change in the next few years?

Mr. T.   By definition, anything we do not understand can't be solid. I have no idea when or how our understanding will change but it will change.

Our last topic will be to examine some of the difficulties of quantum mechanics and relativity. These are very well accepted theories and become very practical in their applications as well as adding greatly to our understanding of the laws of nature. However, even some of our greatest achievements have problems and remain incomplete.

As we try to examine some of the problems, you have to be patient with me because I am not sure I can explain it with the clarity it needs.

J.   Mr. Tweed, if you are going to have trouble from the outset, do you really expect us to understand?

Mr. T.   Some people think that scientists are close to understanding nearly everything. I have a goal that is paradoxical; I want you to have faith in science and I want you know that there are great gaps in our scientific understanding. The gaps represent areas that some very brilliant people are working on. I don't think you can expect me to have an easy time of it.

Problems with quantum mechanics and relativity

Let's start with relativity. The biggest problem is black holes. Inside a black hole, the general theory of relativity predicts the density of matter and the strength of the magnetic field quickly become infinite. Infinities are not physical; they are not part of the real world. We have no verifiable theory that describes the interior (inside the event horizon) of a black hole.

K.   Do you mean that Einstein's theory is wrong?

Mr. T.   No! It just doesn't cover every situation; it does not tell us everything.

K.   So what is wrong with quantum mechanics?

Mr. T.   Let me use the electron as an example. When we see an electron hit something, an examination of the collision indicates that the electron's size is infinitesimal. However, when we use quantum mechanics we must consider the electron's wave properties, and everything becomes diffuse. In the case of the ground state of an atom, the electron can be anywhere within the wave function (orbital) described by the wave equation. We have no idea other than this where it is. Quantum mechanics only deals in probabilities and not specifics. In some situations we have to come to the conclusion that the electron is in more than one place at the same time. As soon as you try to determine a particle's position accurately, you have an uncertainty as to its momentum (uncertainty principle) and therefore its kinetic energy.

Scientists like Bohr were ok with this view. Einstein and others thought that the theory was not able to properly describe reality. The situation became worse when the theories of quantum electrodynamics and quantum fields were added to the mix. These theories were extremely successful in solving many problems and adding to our understanding of the subatomic and atomic worlds. However, they required a great many calculations which sometimes produced infinities. Some very smart people like **Richard Feynman** were able to find ways to compensate for or avoid these infinities, but even he called this a "*dippy*" process. Many hope that some new theory will come along and avoid these difficulties.

Quantum mechanics also has another shortcoming; it does not include gravity. One of the "*holy grails*" of physics is to produce a theory of *quantum gravity*. Some hope that this will remove the infinities produced by the quantum electrodynamics' calculations.

J.   This is our last session. You have gone out of your way to explain some things that scientists don't understand. I am not even able to understand what it is that they don't understand; it's too complicated. Why didn't you just summarize what we learned and try to make us feel like we had accomplished something? What did you expect us to gain from these sessions? What have we accomplished?

Mr. T.   I will enumerate some of the things I expect.

1.   I expect you to have a greater understanding and appreciation of the process of science and scientific discovery.
2.   I expect you to have some feeling for what the scientist calls the laws of nature (physics). You should understand the importance of mathematics in the formulation of these laws even though you may not remember all the details.
3.   You should have gained some perspective of the physical dimensions of our world. We have made calculations involving very small (atoms) and very large (stellar) things. We have also worked with times smaller than picoseconds and larger than 10 billion years.
4.   In our society people have different ideas with regard to the value of science and scientists. On one end of the spectrum of public opinion, some people believe that science has no value and scientists are not to be believed or trusted. On the other end, some people believe that science can or will be able to explain everything. I expect this course to help you determine and justify where you fall in this spectra of public opinion and philosophy.

5.   I expect you will leave these sessions with a number of facts and concepts that are new to you. These are important but only part of the whole process of becoming an educated person.

As to accomplishments:

1.   I hope these sessions have awed you by this process of discovery of the nature of the world around us. We are so often like blind mice feeling the feet of an elephant and trying to determine what kind of animal it is.
2.   I also hope you are able to see and feel the beauty of the structure of our natural laws and descriptions of our world and universe. They are really quite extraordinary.

J.   If we really want to become educated individuals, where do we go from here?

Mr. T.   Anywhere you would like! Have fun!

The End

Appendix I
Big Bang Time Line

The Big Bang
10^{-43} seconds After Big Bang (ABB)

The universe begins with a cataclysm that generates space and time, as well as all the matter and energy the universe will ever hold. For an incomprehensibly small fraction of a second, the universe is an infinitely dense, hot fireball. The prevailing theory describes a peculiar form of energy that can suddenly push out the fabric of space. At 10^{-35} to 10^{-33} seconds a runaway process called "**Inflation**" causes a vast expansion of space filled with this energy. The inflationary period is stopped only when this energy is transformed into matter and energy as we know it.

The Universe Takes Shape
10^{-6} seconds ABB

After inflation, one millionth of a second after the Big Bang, the universe continues to expand but not nearly so quickly. As it expands, it becomes less dense and cools. The most basic forces in nature become distinct: first gravity, then the strong force, which holds nuclei of atoms together, followed by the weak and electromagnetic forces. By the first second, the universe is made up of fundamental particles and energy: quarks, electrons, photons, neutrinos and less familiar types. These particles smash together to form protons and neutrons.

Formation of Basic Elements
3 to 20 minutes ABB

Protons and neutrons come together to form the nuclei of simple elements: hydrogen, helium and lithium.

Recombination
377,000 years ABB

At this time, neutral atoms are formed as electrons link up with hydrogen and helium nuclei. The universe becomes transparent. The microwave background radiation hails from this moment.

Birth of Stars and Galaxies
300 million years ABB

Gravity amplifies slight irregularities in the density of the primordial gas. Even as the universe continues to expand rapidly, pockets of gas become more and more dense. Stars ignite within these pockets, and groups of stars become the earliest galaxies.

Birth of the Sun
~9 Billion years ABB or 5 Billion Years Before the Present (BP)
The sun forms within a cloud of gas in a spiral arm of the Milky Way Galaxy. A vast disk of gas and debris that swirls around this new star gives birth to planets, moons, and asteroids. Earth is the third planet out.

Earliest Life
3.8 Billion Years BP
The Earth has cooled and an atmosphere develops. Microscopic living cells, neither plants nor animals, begin to evolve and flourish in earth's many volcanic environments.

Primitive Animals Appear
700 Million Years BP
These are mostly flatworms, jelly fish and algae. By 570 million years before the present, large numbers of creatures with hard shells suddenly appear.

The First Mammals Appear
200 Million Years BP
The first mammals evolved from a class of reptiles that evolved mammalian traits, such as a segmented jaw and a series of bones that make up the inner ear.

Dinosaurs Become Extinct
65 Million Years BP
An asteroid or comet slams into the northern part of the Yucatan Peninsula in Mexico. This world-wide cataclysm brings to an end the long age of the dinosaurs, and allows mammals to diversify and expand their ranges.

Homo Sapiens Evolve
600,000 Years BP or 13.7 Billion years ABB
Our earliest ancestors evolve in Africa.

The End
10^{100} Years ABB
Galaxies collapse into black holes and black holes evaporate. The universe is dark and dead.

Appendix II
Physical Constants

Quantity	*Symbol*	*Value*
Gravitation constant	G	$6.673 \times 10^{-11} \frac{N \cdot m^2}{kg^2}$
Speed of light in vacuum	c	$3.00 \times 10^8 \frac{m}{s}$
Electron charge	e	$1.60 \times 10^{-19} C$
Plank's constant	h	$6.63 \times 10^{-34} J \cdot s$
Boltsmann's constant	k_B	$1.38 \times 10^{-23} \frac{J}{^\circ K}$
Coulomb constant	k_C	$8.99 \times 10^9 \frac{N \cdot m^2}{C^2}$
Electron mass	m_e	$9.11 \times 10^{-31} kg$
Proton mass	m_p	$1.67265 \times 10^{-27} kg$
Neutron mass	m_n	$1.67495 \times 10^{-27} kg$
Atomic mass unit	*amu*	$1.66 \times 10^{-27} kg$

Appendix III
Useful Information

Solar System

	Mass (kg)	Period (s)	Radius (m)	Dist. from Sun (m)
Mercury	3.18×10^{23}	7.60×10^{6}	2.43×10^{6}	5.79×10^{10}
Venus	4.88×10^{24}	1.94×10^{7}	6.06×10^{6}	1.08×10^{11}
Earth	5.98×10^{24}	3.16×10^{7}	6.37×10^{6}	$1.50 \times 10^{11} m$ (1 au)
Mars	6.42×10^{23}	5.94×10^{7}	3.37×10^{6}	2.28×10^{11}
Jupiter	1.90×10^{27}	3.74×10^{8}	6.99×10^{7}	7.78×10^{11}
Saturn	5.68×10^{26}	9.35×10^{8}	5.85×10^{7}	1.43×10^{12}
Uranus	8.68×10^{25}	2.64×10^{9}	2.33×10^{7}	2.87×10^{12}
Neptune	1.03×10^{26}	5.22×10^{9}	2.21×10^{7}	4.50×10^{12}
Pluto	$\sim 1.4 \times 10^{22}$	7.82×10^{9}	$\sim 1.5 \times 10^{6}$	5.91×10^{12}
Sun	1.99×10^{30}		6.96×10^{8}	0

Surface Temperature = 6000 °C   Interior Temp.=15×10^{6} °C

Moon   Radius = 1.74×10^{6} m Dist. from Earth = 3.84×10^{8} m
            Mass = 7.36×10^{22} kg

Earth's surface gravity acceleration = $9.8 \frac{m}{s^2}$

Atmospheric pressure at sea level = $1.01 \times 10^{5} \frac{N}{m^2}$ (Pa)

Speed of sound (1 Atm, 0 °C) = $343 \frac{m}{s}$

0°C = 273.15 °K

Light year (LY) = $9.48 \times 10^{15} m$

1 au = $1.50 \times 10^{11} m$

$parsec$ = 3.26 LY

Proxima Centauri (nearest star) dist. = 1.31 $parsec$ = 4.27 LY = $4.05 \times 10^{16} m$

Appendix IV
Metric prefix table

prefix	symbol	Power of 10	short scale (UK/US etc)	long scale (Europe exc UK)
deca	da	10^1	ten	ten
hecto	h	10^2	hundred	hundred
kilo	k	10^3	thousand	thousand
mega	M	10^6	million	million
giga	G	10^9	billion	milliard
tera	T	10^{12}	trillion	billion
peta	P	10^{15}	quadrillion	billiard
exa	E	10^{18}	quintillion	trillion
zetta	Z	10^{21}	sextillion	trilliard
yotta	Y	10^{24}	septillion	quadrillion
deci	d	10^{-1}	tenth	tenth
centi	c	10^{-2}	hundredth	hundredth
milli	m	10^{-3}	thousandth	thousandth
micro	μ	10^{-6}	millionth	millionth
nano	n	10^{-9}	billionth	milliardth
pico	p	10^{-12}	trillionth	billionth
femto	f	10^{-15}	quadrillionth	billiardth
atto	a	10^{-18}	quintillionth	trillionth
zepto	z	10^{-21}	sextillionth	trilliardth

www.ingramcontent.com/pod-product-compliance
Lightning Source LLC
Chambersburg PA
CBHW031833170526
45157CB00001B/286